Power Losses in

Switched Inductors

By

Gabriel Alfonso Rincón-Mora

School of Electrical and Computer Engineering
Georgia Institute of Technology

Rincon-Mora.gatech.edu

Contents

List of Figures

List of Abbreviations

BJT $\equiv$ Bipolar-Junction Transistor
CCM $\equiv$ Continuous-Conduction Mode
CMOS $\equiv$ Complementary Metal-Oxide-Semiconductor
DCM $\equiv$ Discontinuous-Conduction Mode
ESD $\equiv$ Electrostatic-Discharge Protection
ESR $\equiv$ Equivalent Series Resistance
FET $\equiv$ Field-Effect Transistor
FM $\equiv$ Frequency Modulation
LED $\equiv$ Light-Emitting Diode
MOS $\equiv$ Metal–Oxide–Semiconductor
MPP $\equiv$ Maximum-Power Point
MPPT $\equiv$ Maximum-Power-Point Tracker
NMOS $\equiv$ N-Channel MOSFET
PDCM $\equiv$ Pseudo Discontinuous-Conduction Mode
PFM $\equiv$ Pulse Frequency Modulation
PMOS $\equiv$ P-Channel MOSFET
RMS $\equiv$ Root–Mean–Square
SL $\equiv$ Switched Inductor
ZCS $\equiv$ Zero-Current Switching
ZVS $\equiv$ Zero-Voltage Switching

C_{CH} $\equiv$ Channel Capacitance
C_{DB} $\equiv$ Drain-Body Capacitance
C_G $\equiv$ Gate Capacitances
C_{GD} $\equiv$ Gate-Drain Capacitance
C_{GS} $\equiv$ Gate-Source Capacitance
C_{OL} $\equiv$ Overlap Capacitance
C_{SB} $\equiv$ Source-Body Capacitance
C_{SWI} $\equiv$ Input Switch-Node Capacitance
C_{SWO} $\equiv$ Output Switch-Node Capacitance

d_D $\equiv$ Drain Duty Cycle
d_{DO} $\equiv$ Output Duty Cycle
d_E $\equiv$ Energize Duty Cycle
d_{EI} $\equiv$ Input Duty Cycle
D_{DG} $\equiv$ Ground Drain Diode
D_{DO} $\equiv$ Output Drain Diode
D_{DT} $\equiv$ Dead-Time Diode

f_{SW} $\equiv$ Switching Frequency

i_D $\equiv$ Driver Current
i_{DS} $\equiv$ Drain-Source Current

$i_G \equiv$ Gate Current
$i_{LD} \equiv$ Load current
$i_{IN} \equiv$ Input Current
$i_L \equiv$ Inductor Current
$i_{L(HI)} \equiv$ Inductor Current's High CCM Peak
$i_{L(LO)} \equiv$ Inductor Current's Low CCM Peak
$i_{L(PK)} \equiv$ Inductor Current's DCM Peak
$i_O \equiv$ Output Current
$i_{OFF} \equiv$ Off Current
$\Delta i_L \equiv$ Inductor Ripple Current
$i_\Delta \equiv$ Triangular Current
$I_S \equiv$ Reverse Saturation Current

$K' \equiv$ Transconductance Parameter

$L_{CH} \equiv$ Channel Length
$L_{MIN} \equiv$ Minimum Allowable Length

$P_B \equiv$ Battery Power
$P_D \equiv$ Drive Power
$P_{DT} \equiv$ Dead-Time Power
$P_G \equiv$ Gate-Charge Power
$P_{GI} \equiv$ Driver Gate-Charge Power
$P_{IN} \equiv$ Input Power
$P_{IV} \equiv i_{DS}$–v_{DS} Overlap Power
$P_{LD} \equiv$ Load Power
$P_{LOSS} \equiv$ Power Losses
$P_{MOS} \equiv$ MOS Power
$P_O \equiv$ Output Power
$P_{OFF} \equiv$ Cut-Off Power
$P_R \equiv$ Ohmic Power
$P_{SWI} \equiv$ Input Switch-Node Power
$P_{SWO} \equiv$ Output Switch-Node Power

$q_{RR} \equiv$ Reverse-Recovery Charge

$R_{CH} \equiv$ Channel Resistance
$R_D \equiv$ Drain Resistance
$R_{DG} \equiv$ Ground Drain Resistance
$R_{DO} \equiv$ Output Drain Resistance
$R_E \equiv$ Energize Resistance
$R_{EG} \equiv$ Ground Energize Resistance
$R_{EI} \equiv$ Input Energize Resistance
$R_{L(AC)} \equiv$ Inductor's AC Resistance
$R_{L(DC)} \equiv$ Inductor's DC Resistance
$R_N \equiv$ Pull-Down N-Type Resistance

$R_{OFF} \equiv$ Off Resistance
$R_P \equiv$ Pull-Up P-Type Resistance
$R_S \equiv$ Source Resistance

$S_{DG} \equiv$ Ground Drain Switch
$S_{DO} \equiv$ Output Drain Switch
$S_{EG} \equiv$ Ground Energize Switch
$S_{EI} \equiv$ Input Energize Switch

$t_C \equiv$ Conduction Time
$t_D \equiv$ Drain Time
$t_{DT} \equiv$ Dead Time
$t_E \equiv$ Energize Time
$t_{SW} \equiv$ Switching Period
$T_J \equiv$ Junction Temperature
$\tau_F \equiv$ Forward Transit Time

$v_B \equiv$ Battery/Battery Voltage
$v_D \equiv$ Drain Voltage
$v_{DD} \equiv$ Power Supply
$v_{DS} \equiv$ Drain–Source Voltage
$v_{DS(SAT)} \equiv$ Saturation Voltage
$v_E \equiv$ Energize Voltage
$v_{GS} \equiv$ Gate–Source Voltage
$v_{IN} \equiv$ Input/Input Voltage
$v_O \equiv$ Output/Output Voltage
$v_S \equiv$ Source/Source Voltage
$v_{SWI} \equiv$ Input Switching Node/Voltage
$v_{SWO} \equiv$ Output Switching Node/Voltage
$v_T \equiv$ MOS Threshold Voltage
$v_{TH} \equiv$ Gate–Source Voltage Threshold
$V_{T0} \equiv$ Zero-Bias Threshold

$W_{CH} \equiv$ Channel Width
$W_{CH}' \equiv$ Optimal Channel Width

$\lambda \equiv$ Channel-Length Modulation Parameter
$\eta_C \equiv$ Power-Conversion Efficiency
$\sigma_{LOSS} \equiv$ Fractional Loss

Power Losses in Switched Inductors

Switched-inductor (SL) power supplies are pervasive in electronic systems because they output a large fraction of the power they draw from their inputs. The main reason for this is the voltages that switches drop are a very small fraction of the output voltage. So the inductor current usually delivers a lot more power into the output than switches consume.

Still, the heat that burning power generates can compromise electronic performance and mechanical integrity. And losing battery energy or ambient power to the switched inductor reduces the charge life or functionality of a system. So understanding the nature, makeup, and sensitivity of these losses is important.

The most fundamental of these is *conduction power*. This is the power that components consume when they conduct inductor current. Series resistances, diodes, and transistors are to blame for this. Another loss is the power that gate drivers need to transition switches between states. Stray capacitances and large switches also leak power.

The operating mechanics of the switched inductor dictate how these components dissipate power. Quantifying losses, however, is not enough. Their significance ultimately rests on the applications they serve.

1. Power Conversion

Power-conversion efficiency η_C is the fraction of *input power* P_{IN} that the *input* v_{IN} delivers to the *output* v_O in Fig. 1:

$$\eta_C \equiv \frac{P_O}{P_{IN}} = \frac{P_O}{P_O + P_{LOSS}} = \frac{P_{IN} - P_{LOSS}}{P_{IN}} = 1 - \frac{P_{LOSS}}{P_{IN}} = 1 - \sigma_{LOSS}. \qquad (1)$$

In addition to this *output power* P_O, P_{IN} also supplies *power losses* P_{LOSS}. So P_O outputs the difference $P_{IN} - P_{LOSS}$, *fractional loss* σ_{LOSS} is the fraction of P_{IN} lost in P_{LOSS}, and η_C is below 100% by the amount σ_{LOSS} sets. η_C and σ_{LOSS} are complementary metrics for transfer efficacy.

Fig. 1. Power supply.

1.1. Voltage Regulators and LED Drivers

Voltage regulators incorporate feedback loops that keep v_O near a prescribed target. This way, v_O hardly varies with the *load current* i_{LD} that v_O in Fig. 2 supplies. The v_O that *light-emitting diode* (LED) *drivers* set is also steady because these drivers keep the *output current* i_O near a user-defined target. Since v_O is steady either way and i_O and i_{LD} are independent variables, engineers normally calculate P_{LOSS} and show how η_C varies across i_O or the P_O that i_O at v_O outputs:

$$\eta_{C(R)} = \frac{P_O}{P_{IN}} = \frac{P_O}{P_O + P_{LOSS}} = \frac{i_{LD} v_O}{i_{LD} v_O + P_{LOSS}} = \frac{i_O v_O}{i_O v_O + P_{LOSS}}. \qquad (2)$$

v_{IN} is usually a good voltage source (with low *source resistance* R_S), so v_{IN} can supply all the P_{IN} that i_O at v_O and P_{LOSS} require.

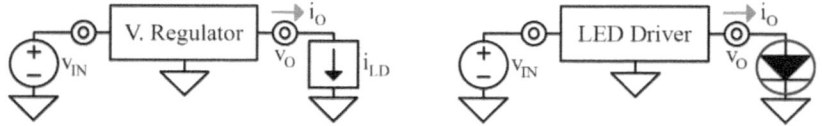

Fig. 2. Voltage regulator and LED driver.

1.2. Battery Chargers

Battery chargers normally incorporate feedback loops that keep i_O in Fig. 3 steady. This i_O charges a *battery* v_B across its operating range. Since i_O is steady and v_B climbs across a prescribed range, showing how η_C varies across v_B is often more revealing than across the P_O that i_O at v_B set:

$$\eta_{C(C)} = \frac{P_O}{P_{IN}} = \frac{P_O}{P_O + P_{LOSS}} = \frac{i_O v_B}{i_O v_B + P_{LOSS}}. \qquad (3)$$

v_{IN} is typically a low-resistance source that can supply all the P_{IN} that i_O at v_B and P_{LOSS} require.

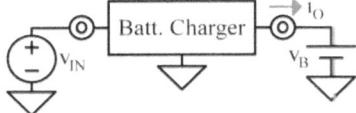

Fig. 3. Battery charger.

1.3. Energy Harvesters

The difference between an ambient-derived *source* v_S and a typical input is that v_S is deficient. In other words, part or all of P_O's range overloads v_S. This is why many *energy harvesters* are chargers that supply what v_S avails. So they cannot always output the i_O (in Fig. 4) that charges v_B quickly or the i_O that maximizes v_B's *capacity*. Still, ambient energy is so pervasive that they can always charge, albeit slowly (with little i_O) and asynchronously (when ambient energy is available).

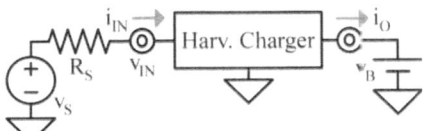

Fig. 4. Energy-harvesting charger.

Smarter energy-harvesting systems charge and supply loads. This is possible when P_{IN}'s maximum exceeds P_O's minimum. So when P_{IN} in Fig. 5 surpasses P_O enough to sustain P_{LOSS}, the harvester supplies P_O and charges v_B with excess P_{IN}. Otherwise, the harvester draws assistance from v_B, in which case P_{IN} and *battery power* P_B supply P_O and P_{LOSS}.

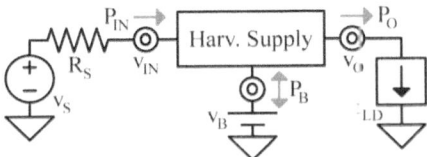

Fig. 5. Energy-harvesting supply.

v_S here is the effective source that *transducers* establish when converting ambient energy into electrical power. R_S models the imperfections that current-limit v_S. v_{IN} therefore peaks at v_S when *input current* i_{IN} is zero and i_{IN} maxes to v_S/R_S when the harvester grounds v_{IN}.

R_S also limits P_{IN}. R_S is also present in conventional regulators, LED drivers, and chargers. In these, however, $P_{IN(MAX)}$ is greater than $P_{O(MAX)}$, so P_O does not overload P_{IN}. Ambient sources, on the other hand, do not always avail the same P_{IN}, so P_O in harvesters can and will at times overload P_{IN}.

A. Maximum-Power Point

The significance of power-conversion efficiency is that reducing losses conserves energy. η_C is less consequential in a harvester because unused ambient energy transforms into forms that the transducer cannot tap. So harvesters should convert and deliver all the power possible.

Good harvesters draw the P_{IN} that supplies the highest P_O. The feedback loops that keep them at the *maximum-power point* (MPP) are *MPP trackers* (MPPT). $P_{O(MAX)}$ is therefore a good metric for harvesting efficacy. Since $P_{O(MAX)}$ reflects how much energy is available, $P_{O(MAX)}$ changes with ambient conditions.

The highest possible P_O results when η_C peaks at the MPP. At this point, the harvester draws the most P_{IN} and loses the least P_{LOSS}. When ambient conditions change, the P_{IN} that corresponds to the new MPP changes. So η_C shifts from its peak and $P_{O(MPP)}$ is no longer $P_{O(MAX)}$.

MPPTs usually keep P_O near $P_{O(MPP)}$ by adjusting P_{IN}. Engineers try to max η_C at the most probable P_{IN} so $P_{O(MPP)}$ matches $P_{O(MAX)}$ more frequently. The general aim is to keep η_C high across P_{IN}'s range.

2. Operating Mechanics

The purpose of a switched inductor is to transfer v_N energy to v_O. For this, *input* and *ground energize switches* S_{EI} and S_{EG} in Fig. 6 energize L_X from v_{IN} and *ground* and *output drain switches* S_{DG} and S_{DO} drain L_X into v_O in alternating phases. This way, v_{IN} produces an *inductor current* i_L that draws P_{IN} from v_{IN} and outputs P_O to v_O.

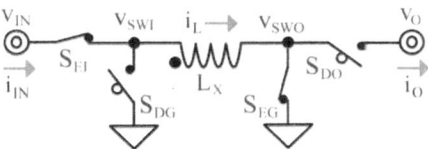

Fig. 6. Switched inductor.

v_{IN} produces an *energize voltage* v_E that raises i_L across *energize time* t_E in Fig. 7. v_O similarly establishes an opposing *drain voltage* v_D that reduces i_L across *drain time* t_D. i_L rises and falls this way across t_C to produce a *inductor ripple current* Δi_L that repeats across cycles:

$$\Delta i_L = \left(\frac{v_E}{L_X}\right) t_E = \left(\frac{v_D}{L_X}\right) t_D. \tag{4}$$

Here, *energize* and *drain duty cycles* d_E and d_D refer to corresponding t_E's and t_D's fractions of t_C: t_E/t_C and t_D/t_C.

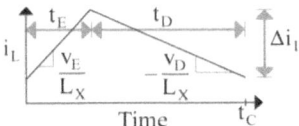

Fig. 7. Inductor current.

So t_E to t_D's ratio follows d_E to d_D's and matches v_D to v_E's:

$$\frac{t_E}{t_D} = \frac{d_E}{d_D} = \frac{v_D}{v_E} = \frac{d_E}{1-d_E}. \tag{5}$$

Since t_D is $t_C - t_E$ and d_D is $1 - d_E$, d_E is v_D's fraction of v_E and v_D:

$$d_E = \frac{v_D}{v_E + v_D}. \tag{6}$$

d_E is therefore a function of the v_E and v_D that v_{IN} and v_O set.

2.1. Continuous Conduction

In *continuous-conduction mode* (CCM), L_X conducts continuously across the entire *switching period* t_{SW}. This way, t_E and t_D in Fig. 8 establish a *conduction time* t_C that extends across t_{SW}. i_L's *low CCM peak* $i_{L(LO)}$ is usually zero or greater, so $i_{L(AVG)}$ is half i_L's ripple Δi_L.

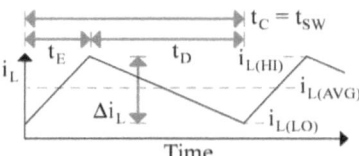

Fig. 8. Inductor current in continuous conduction.

2.2. Discontinuous Conduction

L_X conducts a fraction of t_{SW} in *discontinuous-conduction mode* (DCM). So t_C is less than t_{SW} and i_L in Fig. 9 reaches zero at t_C and remains there until t_{SW} elapses. $i_{L(AVG)}$ across t_{SW} is therefore a fraction of i_L's average $i_{LC(AVG)}$ across t_C, which is half i_L's *DCM peak* $i_{L(PK)}$:

$$i_{L(AVG)} = i_{LC(AVG)}\left(\frac{t_C}{t_{SW}}\right) = \left(\frac{i_{L(PK)}}{2}\right)\left(\frac{t_C}{t_{SW}}\right). \tag{7}$$

$i_{L(PK)}$ is ultimately a reflection of the i_O that $i_{L(AVG)}$ feeds.

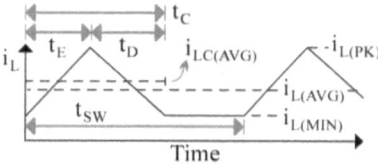

Fig. 9. Inductor current in discontinuous conduction.

Since t_E is a d_E fraction of t_C and v_E across L_X raises i_L to $i_{L(PK)}$ across t_E, t_C also scales with $i_{L(PK)}$:

$$t_C = \frac{t_E}{d_E} = \left(\frac{i_{L(PK)}}{d_E}\right)\left(\frac{L_X}{v_E}\right). \tag{8}$$

i_L averages $0.5i_{L(PK)}$ across t_C and a t_C/t_{SW} fraction of $0.5i_{L(PK)}$ across t_{SW}:

$$i_{L(AVG)} = i_{LC(AVG)}\left(\frac{t_C}{t_{SW}}\right) = \left(\frac{i_{L(PK)}}{2}\right)\left(\frac{t_C}{t_{SW}}\right) = \frac{i_{L(PK)}^2 L_X}{2d_E v_E t_{SW}}. \tag{9}$$

$i_{L(PK)}$ is therefore a squared-root translation of $i_{L(AVG)}$:

$$i_{L(PK)} = \sqrt{2d_E\left(\frac{v_E}{L_X}\right)t_{SW}i_{L(AVG)}} = \sqrt{2d_E t_{SW}\left(\frac{v_E}{L_X}\right)\left(\frac{i_O}{d_{DO}}\right)}, \tag{10}$$

which is a reverse d_{DO} translation of i_O. So t_C and $i_{L(PK)}$ scale with $\sqrt{i_O}$.

Although not often the case, i_L can also fall to and remain at a $i_{L(MIN)}$ that is not zero. In these cases, i_L rises and falls with v_E and v_D across L_X and flattens with zero volts. This is *pseudo DCM* (PDCM).

2.3. Circuit Variants

L_X in Fig. 6 can "step" v_{IN} down or up to a lower or higher v_O. If v_O is less than v_{IN}, $v_{IN} - v_O$ can establish the positive v_E needed to energize L_X. So removing S_{EG} and S_{DO} and connecting L_X to v_O transform the *buck–boost* in Fig. 6 into the *buck* in Fig. 10. In this case, *output duty cycle* d_{DO} is a $t_E + t_D$ fraction of t_C, which is one. So $i_{L(AVG)}$ matches i_O, and since *input duty cycle* d_{EI} is a t_E fraction of t_C, i_{IN} is a d_{EI} fraction of $i_{L(AVG)}$.

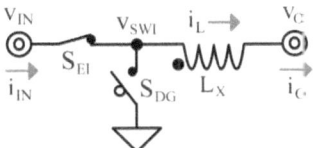

Fig. 10. Switched-inductor buck.

When v_{IN} is less than v_O, $v_O - v_{IN}$ can set the v_D needed to drain L_X. So removing S_{EI} and S_{DG} and connecting v_{IN} to L_X transform the buck–boost into the *boost* in Fig. 11. Here, i_{IN} matches $i_{L(AVG)}$ and i_O is a d_{DO} fraction of $i_{L(AVG)}$. Converting a buck–boost into a boost or a buck when possible is good because fewer switches need less space and power.

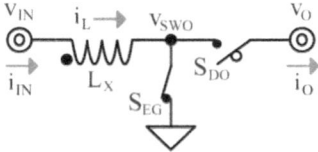

Fig. 11. Switched-inductor boost.

2.4. CMOS Implementation

N-channel metal–oxide–semiconductor (MOS) *field-effect transistors* (FETs) M_{DG} and M_{EG} in Fig. 12 implement the ground switches in Fig. 6 because *P-channel* MOS switches would require negative gate voltages to close. Similarly, PMOS transistors M_{EI} and M_{DO} normally realize the input and output switches because NMOS transistors would require above-v_{IN} and -v_O gate voltages to close. Paralleling an NMOS with M_{EI} or M_{DO} reduces resistance when v_O (or v_{IN}) is much higher than v_{IN} (or v_O), in which case gate voltages can rise well above v_{IN} (or v_O). Note that all source-terminal arrows point in the direction they steer i_L.

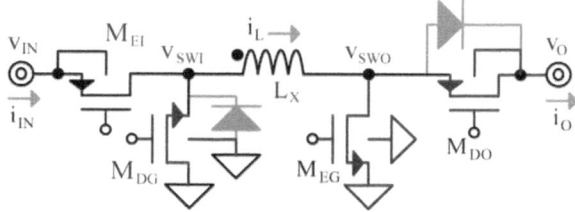

Fig. 12. CMOS switched inductor.

A. Dead-Time Conduction

Dead time t_{DT} between the conduction periods of adjacent switches keep M_{EI}–M_{DG} and M_{EG}–M_{DO} from momentarily grounding v_{IN} and v_O, which would burn too much power. Since v_{IN} directs i_L into v_O, M_{DG}'s and M_{DO}'s body diodes conduct i_L into v_O across these t_{DT}'s. The body connections shown ensure only these body diodes can conduct i_L.

When *MOS threshold voltages* v_T's are less than 500 mV or so, i_L discharges and charges capacitances at the *input* and *output switching nodes* v_{SWI} and v_{SWO} below and above the v_T's needed to engage M_{DG} and M_{DO}. So M_{DG} and M_{DO} conduct all or part of i_L across t_{DT}'s. Still, the effect is similar because M_{DG} and M_{DO} behave like diodes in this mode.

B. Duty Cycle

This diode action ultimately reduces the v_D that drains L_X. But since that happens across two small fractions of t_C and diode voltages are usually small fractions of v_E and v_D, the effect on d_E is typically low. The resulting d_E'' is nevertheless greater than the ideal d_E by these fractions:

$$d_E'' = \frac{v_D}{v_E + v_D} + \left(\frac{v_{DO} + v_{DG}}{v_E + v_D}\right)\left(\frac{2t_{DT}}{t_C}\right) = d_E + \left(\frac{v_{DO} + v_{DG}}{v_E + v_D}\right)\left(\frac{2t_{DT}}{t_C}\right) > d_E. \quad (11)$$

The effect is lower in DCM because the controller opens the drain switches when i_L is zero. So these diodes conduct i_L only across one t_{DT}:

$$d_{E(DCM)}'' = \frac{v_D}{v_E + v_D} + \left(\frac{v_{DO} + v_{DG}}{v_E + v_D}\right)\left(\frac{t_{DT}}{t_C}\right) = d_E + \left(\frac{v_{DO} + v_{DG}}{v_E + v_D}\right)\left(\frac{t_{DT}}{t_C}\right) < d_E''. \quad (12)$$

C. Switching Voltages

When energize switches open, i_L reduces v_{SWI} and raises v_{SWO} until *ground* and *output drain diodes* D_{DG} and D_{DO} conduct i_L. So v_{SWI} falls from v_{IN} to $-v_{DG}$ and v_{SWO} rises from zero to v_{DO} over v_O when t_E ends in Fig. 13. v_{SWI} rises to zero and v_{SWO} falls to v_O a t_{DT} into t_D when S_{DG} and S_{DO} close.

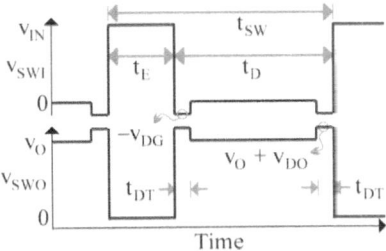

Fig. 13. Switching voltages.

Drain switches S_{DG} and S_{DO} open a t_{DT} before t_D ends, so D_{DG} and D_{DO} pull v_{SWI} a v_{DG} below ground and v_{SWO} a v_{DO} over v_O. And v_{SWI} climbs to v_{IN} and v_{SWO} falls to zero when energize switches S_{EI} and S_{EG} close at the beginning of t_E. This sequence repeats every t_{SW}.

3. Ohmic Loss

Ohmic power P_R refers to power that resistances in the switched inductor consume when conducting i_L. This is a loss because they burn v_{IN} power that v_O does not receive. The average power a device that conducts i_A and drops v_A consumes across time t_X is

$$P_{AX} \equiv P_{A(AVG)}\Big|_0^{t_X} = \frac{1}{t_X} \int_0^{t_X} P_A dt = \frac{1}{t_X} \int_0^{t_X} i_A v_A dt . \tag{13}$$

3.1. Ohmic Power

Since the voltage v_R across a resistor R_X that conducts i_X is $i_X R_X$, R_X power P_R's $i_X v_R$ is $i_X^2 R_X$. When i_X ramps linearly across time t_X like Fig. 14 shows, P_R climbs quadratically with i_X and P_R's average P_{RX} rises quadratically with i_X's *root–mean–square* (RMS) $i_{X(RMS)}$:

$$P_{RX} = \frac{1}{t_X} \int_0^{t_X} i_X v_R dt = \left(\frac{1}{t_X} \int_0^{t_X} i_X^2 dt \right) R_X = i_{X(RMS)}^2 R_X . \tag{14}$$

This means that P_{RX}'s rise accelerates with i_X.

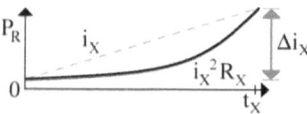

Fig. 14. Resistor power with ramp current.

A. Triangular Current

The *triangular current* i_A in Fig. 15 ramps across t_X to Δi_A. Squaring this i_A and averaging i_A^2 across t_X reduces $i_{A(RMS)}$ to

$$i_{\Delta(RMS)} = \sqrt{\left(\frac{1}{t_X}\right) \int_0^{t_X} \left(\frac{\Delta i_\Delta t}{t_X}\right)^2 dt} = \sqrt{\left(\frac{\Delta i_\Delta^2}{t_X^3}\right)\left(\frac{t_X^3}{3}\right)} = \frac{\Delta i_\Delta}{\sqrt{3}}. \qquad (15)$$

So the RMS of a triangular current is a root-three fraction of its peak Δi_Δ.

Fig. 15. Triangular current.

B. Alternating Current

Positive and negative currents through a resistor burn similar power. RMS accounts for this because its squaring function desensitizes RMS from polarity. So when an *alternating current* i_{AC} is symmetrical, i_{AC}'s negative half burns as much power as i_{AC}'s positive half.

i_{AC} across each half in Fig. 16 is triangular like i_Δ in Fig. 15. So $i_{AC(RMS)}$ across each $0.5t_X$ is the same across t_X and like i_Δ across $0.5t_X$:

$$i_{AC(RMS)} \equiv i_{AC(RMS)}\Big|_{t_X} = i_{AC(RMS)}\Big|_{0.5t_X} = i_{\Delta(RMS)}\Big|_{0.5t_X} = \frac{0.5\Delta i_{AC}}{\sqrt{3}}. \qquad (16)$$

And since each half triangle traverses $0.5\Delta i_{AC}$, $i_{AC(RMS)}$ is a root-three fraction of half i_{AC}'s ripple Δi_{AC}.

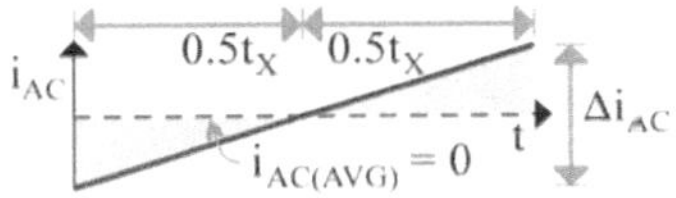

Fig. 16. Alternating triangular current.

C. Power Theorem

i_X in Fig. 17 ramps about an $i_{X(AVG)}$ that is greater than zero. Since i_X rises from $i_{X(MIN)}$ across Δi_X in t_X and $i_{X(MIN)}$ is half a ripple below i_X's average, i_X across time t is

$$i_X = i_{X(MIN)} + \left(\frac{\Delta i_X}{t_X}\right)t = i_{X(AVG)} - \frac{\Delta i_X}{2} + \left(\frac{\Delta i_X}{t_X}\right)t. \qquad (17)$$

Squaring i_X and averaging $i_X{}^2$ across t_X reveals that $i_{X(RMS)}{}^2$ decomposes into average and alternating components $i_{X(AVG)}{}^2$ and $i_{AC(RMS)}{}^2$:

$$i_{X(RMS)}{}^2 = \frac{1}{t_X} \int_0^{t_X} i_X{}^2 dt$$

$$= \frac{1}{t_X}\left[i_{X(MIN)}{}^2 t + \left(\frac{\Delta i_X{}^2}{t_X{}^2}\right)\left(\frac{t^3}{3}\right) + 2i_{X(MIN)}\left(\frac{\Delta i_X}{t_X}\right)\left(\frac{t^2}{2}\right)\right]\Bigg|_0^{t_X}$$

$$= \frac{1}{t_X}\left[\left(i_{X(AVG)} - \frac{\Delta i_X}{2}\right)^2 t_X + \frac{\Delta i_X{}^2 t_X}{3} + \left(i_{X(AVG)} - \frac{\Delta i_X}{2}\right)\Delta i_X t_X\right]. \quad (18)$$

$$= i_{X(AVG)}{}^2 + \left(\frac{\Delta i_X}{2}\right)^2\left(1 + \frac{4}{3} - 2\right) = i_{X(AVG)}{}^2 + \left(\frac{0.5\Delta i_X}{\sqrt{3}}\right)^2$$

$$= i_{X(AVG)}{}^2 + i_{AC(RMS)}{}^2$$

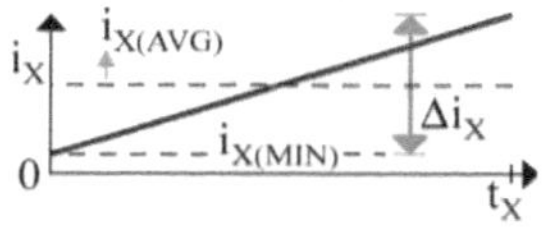

Fig. 17. Non-zero crossing ramp current.

When R_X conducts i_X across a t_X fraction of the switching period, R_X consumes a similar t_{SW} fraction of P_{RX}. So P_R reduces to

$$P_R = P_{RX}\left(\frac{t_X}{t_{SW}}\right) = i_{X(RMS)}{}^2 R_X\left(\frac{t_X}{t_{SW}}\right) = \left(i_{X(AVG)}{}^2 + i_{AC(RMS)}{}^2\right)R_X\left(\frac{t_X}{t_{SW}}\right). \quad (19)$$

This expression is very useful because it extrapolates the power that resistances burn when conducting irregular fractions of t_{SW}.

3.2. Continuous Conduction

A. Switched Inductor

i_L in CCM ripples about an $i_{L(AVG)}$ that keeps $i_{L(LO)}$ at or above zero. L_X's *equivalent series resistance* (ESR) R_L conducts this i_L across all of t_{SW}. But since the *skin effect* keeps dynamic current near the edges of the coil, R_L is higher for the ripple than for the static part of i_L. So $i_{L(AVG)}$ mostly burns power with the *inductor's dc resistance* $R_{L(DC)}$ and $i_{AC(RMS)}$ with the *inductor's* higher *ac resistance* $R_{L(AC)}$:

$$P_{RL} \approx i_{L(AVG)}{}^2 R_{L(DC)} + i_{AC(RMS)}{}^2 R_{L(AC)}$$

$$= \left(\frac{i_O}{d_{DO}} \right)^2 R_{L(DC)} + \left(\frac{0.5\Delta i_L}{\sqrt{3}} \right)^2 R_{L(AC)} \qquad (20)$$

where $i_{L(AVG)}$ is a reverse d_{DO} translation of i_O and $i_{AC(RMS)}$ is a root-three fraction of half Δi_L.

Example 1: Determine P_{RL} and σ_{RL} for L_X in an ideal buck–boost in CCM when v_{IN} is 2 V, v_O is 4 V, i_O is 250 mA, L_X is 10 µH, t_{SW} is 1 µs, and R_L is 200 mΩ.

Solution:

$$d_{EI} = d_E \approx \frac{v_O}{v_{IN} + v_O} = \frac{4}{2+4} = 67\%$$

$$\therefore \quad d_{DO} = d_D = 1 - d_E = 1 - 67\% = 33\%$$

$$\Delta i_L = \left(\frac{v_E}{L_X} \right) d_E t_{SW} = \left(\frac{2}{10\mu} \right)(67\%)(1\mu) = 130 \text{ mA}$$

$$P_{RL} \approx \left[\left(\frac{i_O}{d_{DO}} \right)^2 + \left(\frac{0.5\Delta i_L}{\sqrt{3}} \right)^2 \right] R_L$$

$$= \left\{ \left(\frac{250\text{m}}{33\%} \right)^2 + \left[\frac{0.5(130\text{m})}{\sqrt{3}} \right]^2 \right\} (200\text{m}) = 120 \text{ mW}$$

$$\sigma_{RL} = \frac{P_{RL}}{(i_O/d_{DO})d_{EI}v_{IN}} \approx \frac{120\text{m}}{(250\text{m}/33\%)(67\%)(2)} = 12\%$$

Note: R_L consumes roughly 12% of P_{IN}.

B. Energize and Drain Resistances

Energize switches conduct i_L only across t_E. Their current i_{RE} is therefore i_L across t_E and zero across t_D, like Fig. 18 shows. *Energize resistance* R_E therefore consumes a t_E/t_{SW} fraction of the power R_E burns across t_E. Since i_{RE}'s average and ripple across t_E match i_L's $i_{L(AVG)}$ and Δi_L, *input and ground energize resistances* R_{EI} and R_{EG} in R_E dissipate

$$P_{RE} = i_{RE(RMS)}{}^2 R_E$$

$$= \left(i_{E(AVG)}{}^2 + i_{AC(RMS)}{}^2 \right) R_E \left(\frac{t_E}{t_{SW}} \right) \qquad , \qquad (21)$$

$$= \left[\left(\frac{i_O}{d_{DO}} \right)^2 + \left(\frac{0.5 \Delta i_L}{\sqrt{3}} \right)^2 \right] \left(R_{EI} + R_{EG} \right) d_E$$

where i_{RE}'s $i_{E(AVG)}$ and $i_{AC(RMS)}$ across t_E match i_L's across t_{SW}.

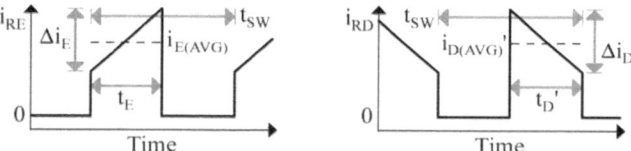

Fig. 18. Energize and drain resistor currents.

Drain components similarly conduct i_L only across t_D. Dead times, however, shorten the times that drain switches close. So drain-switch current i_{RD} is i_L across t_D' or $t_D - 2t_{DT}$ and zero across t_E. *Drain resistance* R_D therefore consumes a t_D'/t_{SW} fraction of the power R_D burns across t_D'. Since i_{RD}'s average across t_D' matches i_L's $i_{L(AVG)}$ when t_{DT}'s are symmetrical and i_{RD}'s ripple matches i_L's Δi_L when t_{DT}'s are much shorter than t_{SW}, *ground* and *output drain resistances* R_{DG} and R_{DO} in R_D burn

$$P_{RD} = i_{RD(RMS)}{}^2 R_D$$

$$= \left(i_{D(AVG)}{}^2 + i_{AC(RMS)}{}^2 \right) R_D \left(\frac{t_D - 2t_{DT}}{t_{SW}} \right)$$

$$\approx \left[\left(\frac{i_O}{d_{DO}} \right)^2 + \left(\frac{0.5\Delta i_L}{\sqrt{3}} \right)^2 \right] (R_{DG} + R_{DO}) \left(d_D - \frac{2t_{DT}}{t_{SW}} \right), \quad (22)$$

where i_{RD}'s $i_{E(AVG)}$ and $i_{AC(RMS)}$ across t_D roughly match i_L's across t_{SW}.

Example 2: Determine P_{RE} and σ_{RE} for M_{EG} in the ideal buck–boost from Example 1 in CCM when R_{EG} is 200 mΩ.

Solution:

$$d_{EI} = d_E = 67\%, \ d_{DO} = d_D = 33\%, \ \Delta i_L = 130 \text{ mA}$$

from Example 1

$$P_{RE} = \left[\left(\frac{i_O}{d_{DO}} \right)^2 + \left(\frac{0.5\Delta i_L}{\sqrt{3}} \right)^2 \right] R_{EG} d_E$$

$$= \left\{ \left(\frac{250m}{33\%} \right)^2 + \left[\frac{0.5(130m)}{\sqrt{3}} \right]^2 \right\} (200m)(67\%) = 77 \text{ mW}$$

$$\sigma_{RE} = \frac{P_{RE}}{(i_O/d_{DO}) d_{EI} v_{IN}} = \frac{77m}{(250m/33\%)(67\%)(2)} = 7.6\%$$

Note: M_{EG} dissipates less of P_{IN} than R_L because M_{EG} conducts i_L a t_E fraction of t_{SW}. But since M_{DC} conducts the other t_D fraction of t_{SW}, M_{EG} and M_{DO} can ccnsume as much as R_L.

C. Output Capacitor

The ultimate aim of voltage regulators and LED drivers is to supply i_O. Supplying this i_O, however, is impossible when the output switch is open. The purpose of C_O in Fig. 19 is to supply i_O when S_{DO} opens.

C_O's dc current must be zero for v_O and i_O to remain steady. So on average, S_{DO} outputs the i_O that feeds the load. But since S_{DO} opens a t_E fraction of t_{SW}, i_{DO}'s $i_{L(AVG)}$ supplies a correspondingly higher d_{DO}

translation of i_O: i_O/d_{DO}. i_{DO} is zero otherwise like Fig. 20 shows. So C_O supplies i_O when S_{DO} opens and receives i_O/d_{DO} minus i_O otherwise. C_O's ESR R_C therefore burns power with i_O across t_E and with i_{DO} across t_{DO}:

$$P_{RC} = P_{RC}\big|_{t_E} + P_{RC}\big|_{t_{DO}}$$

$$= i_O{}^2 R_C\left(\frac{t_E}{t_{SW}}\right) + \left[\left(i_{DO(AVG)} - i_O\right)^2 + i_{AC(RMS)}{}^2\right]R_C\left(\frac{t_{DO}}{t_{SW}}\right), \qquad (23)$$

$$= i_O{}^2 R_C d_E + \left[\left(\frac{i_O}{d_{DO}} - i_O\right)^2 + \left(\frac{0.5\Delta i_L}{\sqrt{3}}\right)^2\right]R_C d_{DO}$$

where R_C's $i_{C(AVG)}$ across t_{DO} is $i_{DO(AVG)}$'s i_O/d_{DO} minus i_O and $i_{AC(RMS)}$ is a root-three fraction of half i_{DO}'s ripple Δi_L.

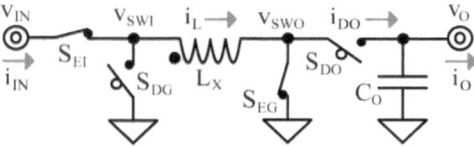

Fig. 19. Switched-inductor voltage regulator or LED driver.

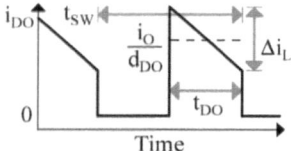

Fig. 20. Duty-cycled inductor drain current.

Example 3: Determine P_{RC} and σ_{RC} for C_O in an ideal boost in CCM when v_{IN} is 2 V, v_O is 4 V, i_O is 250 mA, L_X is 10 μH, t_{SW} is 1 μs, and R_C is 200 mΩ.

Solution:

$$d_E = \frac{v_O - v_{IN}}{v_O} = \frac{4-2}{4} = 50\%$$

$$\therefore \quad d_{DO} = d_D = 1 - d_E = 1 - 50\% = 50\%$$

$$\Delta i_L = \left(\frac{v_E}{L_X}\right) d_E t_{SW} = \left(\frac{2}{10\mu}\right)(50\%)(1\mu) = 100 \text{ mA}$$

$$P_{RC} = i_O{}^2 R_C d_E + \left[\left(\frac{i_O}{d_{DO}} - i_O\right)^2 + \left(\frac{0.5\Delta i_L}{\sqrt{3}}\right)^2\right] R_C d_{DO}$$

$$= (250\text{m})^2 (200\text{m})(50\%)$$

$$+ \left\{\left(\frac{250\text{m}}{50\%} - 250\text{m}\right)^2 + \left[\frac{0.5(100\text{m})}{\sqrt{3}}\right]^2\right\}(200\text{m})(50\%)$$

$$= 13 \text{ mW}$$

$$\sigma_{RC} = \frac{P_{RC}}{\left(i_O/d_{DO}\right) d_{EI} v_{IN}} = \frac{13\text{m}}{(250\text{m}/50\%)(1)(2)} = 1.3\%$$

Note: R_C consumes 1.3% of P_{IN}.

In a buck, L_X connects to v_O. i_L therefore ripples about the average that supplies i_O. So the purpose of C_O in the buck is to supply and sink half i_L's ripple. Since $0.5\Delta i_L$ is oftentimes much lower than the i_O that C_O in Fig. 19 supplies across t_E, P_{RC} in the buck in Fig. 21 is usually lower:

$$P_{RC(BK)} = i_{C(RMS)}{}^2 R_C = i_{AC(RMS)}{}^2 R_C = \left(\frac{0.5\Delta i_L}{\sqrt{3}}\right)^2 R_C, \qquad (24)$$

where i_C's $i_{C(AVG)}$ is zero and $i_{AC(RMS)}$ matches i_L's.

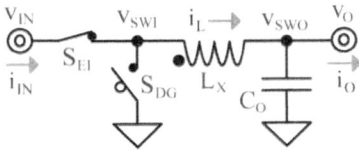

Fig. 21. Switched-inductor buck voltage regulator or LED driver.

Example 4: Determine P_{RC} and σ_{RC} for C_O in an ideal buck in CCM when v_{IN} is 4 V, v_O is 2 V, i_O is 250 mA, L_X is 10 μH, t_{SW} is 1 μs, and R_C is 200 mΩ.

Solution:

$$d_{EI} = d_E \approx \frac{v_O}{v_{IN}} = \frac{2}{4} = 50\% \quad \therefore \quad d_D = 1 - d_E = 1 - 50\% = 50\%$$

$$\Delta i_L = \left(\frac{v_{IN} - v_O}{L_X}\right)d_E t_{SW} = \left(\frac{4-2}{10\mu}\right)(50\%)(1\mu) = 100 \text{ mA}$$

$$P_{RC} = \left(\frac{0.5\Delta i_L}{\sqrt{3}}\right)^2 R_C = \left[\frac{0.5(100m)}{\sqrt{3}}\right]^2 (200m) = 170 \text{ μW}$$

$$\sigma_{RC} = \frac{P_{RC}}{(i_O/d_{DO})d_{EI}v_{IN}} = \frac{170\mu}{(250m/1)(50\%)(4)} < 0.1\%$$

Note: R_C dissipates less of P_{IN} in the buck because R_C only conducts Δi_L. In the boost, R_C conducts i_O-level currents.

3.3. Discontinuous Conduction

A. Switched Inductor

i_L in DCM rises to $i_{L(PK)}$ after t_E and falls to zero before t_{SW} ends. Since no part of i_L is steady across t_{SW}, all of i_L flows through the skin of L_X. This means that $R_{L(AC)}$ is the only part of L_X that consumes P_{RL}. P_{RL} is ultimately a t_C/t_{SW} fraction of the RMS power burned across t_C:

$$P_{RL} = i_{L(RMS)}{}^2 R_{L(AC)} = i_{LC(RMS)}{}^2 R_{L(AC)}\left(\frac{t_C}{t_{SW}}\right) = \left(\frac{i_{L(PK)}}{\sqrt{3}}\right)^2 R_{L(AC)}\left(\frac{t_C}{t_{SW}}\right), \quad (25)$$

where $i_{LC(RMS)}$ is i_L's RMS across t_C, which is a root-three fraction of half i_L's DCM peak $i_{L(PK)}$.

Example 5: Determine P_{RL} and σ_{RL} for L_X in the ideal boost from Example 3 in DCM when i_O is 10 mA and $R_{L(AC)}$ is 200 mΩ.

Solution:

$$d_E = 50\% \text{ and } d_{DO} = d_D = 50\% \text{ from Example 3}$$

$$i_{L(PK)} = \sqrt{2d_E t_{SW}\left(\frac{v_E}{L_X}\right)\left(\frac{i_O}{d_{DO}}\right)}$$

$$= \sqrt{2(50\%)(1\mu)\left(\frac{2}{10\mu}\right)\left(\frac{10m}{50\%}\right)} = 63 \text{ mA}$$

$$t_C = \left(\frac{i_{L(PK)}}{d_E}\right)\left(\frac{L_X}{v_E}\right) = \left(\frac{63m}{50\%}\right)\left(\frac{10\mu}{2}\right) = 630 \text{ ns}$$

$$P_{RL} = \left(\frac{i_{L(PK)}}{\sqrt{3}}\right)^2 R_{L(AC)}\left(\frac{t_C}{t_{SW}}\right)$$

$$= \left(\frac{63m}{\sqrt{3}}\right)^2 (200m)\left(\frac{630n}{1\mu}\right) = 170 \text{ }\mu W$$

$$\sigma_{RL} = \frac{P_{RL}}{\left(i_O/d_{DO}\right)d_{EI}v_{IN}} = \frac{170\mu}{(10m/50\%)(1)(2)} = 0.4\%$$

Note: R_L dissipates 0.4% of P_{IN}.

B. Energize and Drain Resistances

Energize and drain switches similarly consume t_E/t_{SW} and t_D/t_{SW} fractions of the RMS power R_E and R_D burn across t_E and t_D:

$$P_{RE} = i_{RE(RMS)}^2 R_E = i_{E(RMS)}^2 R_E\left(\frac{t_E}{t_{SW}}\right) = \left(\frac{i_{L(PK)}}{\sqrt{3}}\right)^2 (R_{EI} + R_{EG})\left(\frac{t_E}{t_{SW}}\right) \quad (26)$$

$$P_{RD} = i_{RD(RMS)}^2 R_D = i_{D(RMS)}^2 R_D\left(\frac{t_D}{t_{SW}}\right) = \left(\frac{i_{L(PK)}}{\sqrt{3}}\right)^2 (R_{DG} + R_{DO})\left(\frac{t_D}{t_{SW}}\right). \quad (27)$$

R_{EI} and R_{EG} in R_E burn P_{RE} and R_{DG} and R_{DO} in R_D burn P_{RD}. i_L's RMS across t_E, t_D, and t_C are all root-three fractions of $i_{L(PK)}$ because i_L is triangular and peaks at $i_{L(PK)}$ in all three cases.

C. Output Capacitor

In DCM, S_{DO} opens across t_E and the period that follows t_C before t_{SW} ends. This corresponds to the part of t_{SW} that excludes t_{DO}. C_O supplies i_O across this time and the part of i_O that i_L does not when S_{DO} closes. So R_C consumes power with i_O across $t_{SW} - t_{DO}$ and with $i_{DO} - i_O$ across t_{DO}:

$$P_{RC} = P_{RC}\big|_{t_{SW} - t_{DO}} + P_{RC}\big|_{t_{DO}}$$

$$= i_O{}^2 R_C \left(\frac{t_{SW} - t_{DO}}{t_{SW}} \right) + \left[\left(i_{DO(AVG)} - i_O \right)^2 + i_{AC(RMS)}{}^2 \right] R_C \left(\frac{t_{DO}}{t_{SW}} \right). \quad (28)$$

$$= i_O{}^2 R_C \left[1 - d_{DO} \left(\frac{t_C}{t_{SW}} \right) \right] + \left[\left(\frac{i_{L(PK)}}{2} - i_O \right)^2 + \left(\frac{i_{L(PK)}}{2\sqrt{3}} \right)^2 \right] R_C d_{DO} \left(\frac{t_C}{t_{SW}} \right)$$

Across t_{DO}, i_{DO}'s average $0.5i_{L(PK)}$ minus i_O burns steady power and i_{DO}'s ripple $0.5i_{L(PK)}$ burns alternating power.

Example 6: Determine P_{RC} and σ_{RC} for C_O in the ideal boost from Examples 3 and 5 in DCM when R_C is 200 mΩ.

Solution:

$$d_E = 50\%, \; d_{DO} = d_D = 50\%, \; i_{L(PK)} = 63 \text{ mA}, \; t_C = 630 \text{ ns}$$

from Examples 3 and 5

$$P_{RC} = i_O{}^2 R_C \left[1 - d_{DO} \left(\frac{t_C}{t_{SW}} \right) \right]$$

$$+ \left[\left(\frac{i_{L(PK)}}{2} - i_O \right)^2 + \left(\frac{i_{L(PK)}}{2\sqrt{3}} \right)^2 \right] R_C d_{DO} \left(\frac{t_C}{t_{SW}} \right)$$

$$= (10m)^2 (200m) \left[1 - 50\% \left(\frac{630n}{1\mu} \right) \right]$$

$$+ \left[\left(\frac{63m}{2} - 10m \right)^2 + \left(\frac{63m}{2\sqrt{3}} \right)^2 \right] (200m)(50\%) \left(\frac{630n}{1\mu} \right)$$

$$= 64 \ \mu W$$

Note: R_C dissipates less of P_{IN} than R_L in Example 5 because R_C does not conduct the part of i_L that feeds i_O.

In a buck, R_O conducts $i_L - i_O$ across t_{SW} because L_X connects to v_O. But since i_L is non-zero only across t_C, R_C consumes power with $i_L - i_O$ across t_C and with i_O across $t_{SW} - t_C$:

$$P_{RC(BK)} = P_{RC}\big|_{t_C} + P_{RC}\big|_{t_{SW} - t_C}$$

$$= \left[\left(i_{LC(AVG)} - i_O \right)^2 + i_{AC(RMS)}^2 \right] R_C \left(\frac{t_C}{t_{SW}} \right) + i_O^2 R_C \left(\frac{t_{SW} - t_C}{t_{SW}} \right) . \quad (29)$$

$$= \left[\left(\frac{i_{L(PK)}}{2} - i_O \right)^2 + \left(\frac{0.5 i_{L(PK)}}{\sqrt{3}} \right)^2 \right] R_C \left(\frac{t_C}{t_{SW}} \right) + i_O^2 R_C \left(1 - \frac{t_C}{t_{SW}} \right)$$

Across t_C, i_L's average $0.5 i_{L(PK)}$ minus i_O burns steady power and i_L's ripple $0.5 i_{L(PK)}$ burns alternating power.

4. Dead-Time Loss

4.1. Conduction Power

A static dc voltage v_X that conducts a ramping current like i_X in Fig. 17 burns the power that i_X's average across that time $i_{X(AVG)}$ into v_X sets:

$$P_{VX} = \frac{1}{t_X} \int_0^{t_X} i_X v_X dt = \left(\frac{1}{t_X} \int_0^{t_X} i_X dt \right) v_X = i_{X(AVG)} v_X . \quad (30)$$

This $i_{X(AVG)}$ is i_X's minimum $i_{X(MIN)}$ plus half i_X's variation Δi_X:

$$i_{X(AVG)} = \frac{1}{t_X}\int_0^{t_x} i_X dt = \frac{1}{t_X}\int_0^{t_x}\left[i_{X(MIN)} + \left(\frac{\Delta i_X}{t_X}\right)t\right]dt = i_{X(MIN)} + \frac{\Delta i_X}{2}. \qquad (31)$$

So P_V's average P_{VX} climbs linearly with i_X's average $i_{X(AVG)}$.

4.2. Continuous Conduction

Dead-time diodes conduct across t_{DT}'s the i_L that L_X holds before and after L_X energizes. Although i_L ramps over time, t_{DT}'s are usually so much shorter than t_{SW} that i_L is fairly steady across t_{DT}'s. So across the t_{DT} that follows t_E, when L_X begins to drain, i_L is nearly steady at $i_{L(HI)}$. i_L is similarly steady at $i_{L(LO)}$ across the t_{DT} that precedes t_E.

D_{DG} and D_{DO} consume power with i_L's *high CCM peak* $i_{L(HI)}$ and $i_{L(LO)}$ across these t_{DT}'s. Since i_L is steady and diode voltages are fairly insensitive to current variations, these diodes burn static power across t_{DT}'s and dead-time fractions across t_{SW}. So *dead-time power* P_{DT} is

$$P_{DT} \approx i_{L(LO)}\left(v_{DG} + v_{DO}\right)\left(\frac{t_{DT}}{t_{SW}}\right) + i_{L(HI)}\left(v_{DG} + v_{DO}\right)\left(\frac{t_{DT}}{t_{SW}}\right)$$

$$\approx 2i_{L(AVG)}\left(v_{DG} + v_{DO}\right)\left(\frac{t_{DT}}{t_{SW}}\right)$$

$$= \left(\frac{i_O}{d_{DO}}\right)(v_{DG} + v_{DO})\left(\frac{2t_{DT}}{t_{SW}}\right). \qquad (32)$$

Diode voltages can be so insensitive to current variations that D_{DG} and D_{DO} can drop similar v_{DG}'s and v_{DO}'s with $i_{L(HI)}$ and $i_{L(LO)}$. $i_{L(HI)}$ and $i_{L(LO)}$ therefore burn power with approximately equal t_{DT} fractions of the same v_{DG} and v_{DO}. And since $i_{L(HI)}$ and $i_{L(LO)}$ average $i_{L(AVG)}$, $i_{L(HI)}$ and $i_{L(LO)}$ add to $2i_{L(AVG)}$. This means that a reverse d_{DO} translation of i_O burns P_{DT} with v_{DG} and v_{DO} across two t_{DT} fractions of t_{SW}.

Interestingly, P_{DT} is a nearly constant fraction of P_{IN}. This is because P_{DT} and P_{IN} both scale linearly with $i_{L(AVG)}$:

$$\sigma_{DT} = \frac{P_{DT}}{P_{IN}} = \frac{P_{DT}}{i_{IN} v_{IN}} = \frac{P_{DT}}{i_{L(AVG)} d_{EI} v_{IN}} = \left(\frac{v_{DG} + v_{DO}}{d_{EI} v_{IN}} \right) \left(\frac{2t_{DT}}{t_{SW}} \right). \tag{33}$$

So P_{DT}'s fraction of P_{IN} ultimately hinges on the voltage and time fractions that diode voltages and t_{DT}'s establish with $d_{EI} v_{IN}$ and t_{SW}.

Example 7: Determine P_{DT} and σ_{DT} in an ideal boost in CCM when v_{IN} is 2 V, v_O is 4 V, i_O is 250 mA, L_X is 10 μH, t_{SW} is 1 μs, v_{DO} is 800 mV, and t_{DT} is 50 ns.

Solution:

$$d_E = \frac{v_O - v_{IN}}{v_O} + \left(\frac{v_{DO}}{v_O} \right) \left(\frac{2t_{DT}}{t_C} \right)$$

$$= \frac{4-2}{4} + \left(\frac{800m}{4} \right) \left[\frac{2(50n)}{1\mu} \right] = 52\%$$

$$\therefore \quad d_{DO} = d_D = 1 - d_E = 1 - 50\% = 43\%$$

$$\Delta i_L = \left(\frac{v_E}{L_X} \right) d_E t_{SW} \approx \left(\frac{2}{10\mu} \right)(52\%)(1\mu) = 100 \ mA$$

$$P_{DT} \approx \left(\frac{i_O}{d_{DO}} \right) v_{DO} \left(\frac{2t_{DT}}{t_{SW}} \right)$$

$$= \left(\frac{250m}{48\%} \right)(800m) \left[\frac{2(50n)}{1\mu} \right] = 42 \ mW$$

$$\sigma_{DT} = \frac{P_{DT}}{\left(i_O / d_{DO} \right) d_{EI} v_{IN}} \approx \frac{42m}{(250m/48\%)(1)(2)} = 4.0\%$$

Note: Dead-time diodes can consume considerable power.

4.3. Discontinuous Conduction

i_L in discontinuous conduction rises to $i_{L(PK)}$ after t_E and falls to zero across t_D before t_{SW} ends. So i_L is nearly $i_{L(PK)}$ across the first t_{DT} in t_D and nearly zero across the second. Dead-time diodes therefore consume noticeable power only across the first t_{DT}:

$$P_{DT} \approx i_{L(PK)} \left(V_{DG} + V_{DO} \right) \left(\frac{t_{DT}}{t_{SW}} \right). \tag{34}$$

P_{DT}'s fraction of P_{IN} is the product of three ratios:

$$\sigma_{DT} = \frac{P_{DT}}{P_{IN}} = \frac{P_{DT}}{i_{IN} V_{IN}} = \frac{P_{DT}}{i_{L(AVG)} d_{EI} V_{IN}} = \left(\frac{i_{L(PK)}}{i_{L(AVG)}} \right) \left(\frac{V_{DG} + V_{DO}}{d_{EI} V_{IN}} \right) \left(\frac{t_{DT}}{t_{SW}} \right). \tag{35}$$

The current ratio is greater than one, the voltage ratio is nearly one or lower, and the time ratio is usually a small fraction, so diode and t_{DT} fractions counter the effect of current gain. And since $i_{L(AVG)}$ scales faster with i_O than $i_{L(PK)}$ with $\sqrt{i_O}$, P_{DT}'s fraction of P_{IN} falls with increasing $\sqrt{i_O}$.

Example 8: Determine P_{DT} and σ_{DT} in an ideal buck in DCM when v_{IN} is 4 V, v_O is 2 V, i_O is 10 mA, L_X is 10 μH, t_{SW} is 1 μs, t_{DT} is 50 ns, and D_{DG} drops 700 mV.

Solution:

$$d_E = \frac{v_O}{v_{IN}} + \left(\frac{v_{DG}}{v_{IN}} \right) \left(\frac{t_{DT}}{t_C} \right) = \frac{2}{4} + \left(\frac{700m}{4} \right) \left(\frac{50n}{1\mu} \right) = 51\%$$

$$\therefore \quad d_D = 1 - d_E = 1 - 51\% = 49\%$$

$$i_{L(PK)} = \sqrt{2 d_E t_{SW} \left(\frac{v_E}{L_X} \right) i_O}$$

$$= \sqrt{2(51\%)(1\mu)\left(\frac{4-2}{10\mu} \right)(10m)} = 45 \ \text{mA}$$

$$P_{DT} \approx i_{L(PK)} V_{DG} \left(\frac{t_{DT}}{t_{SW}} \right) = (45m)(700m)\left(\frac{50n}{1\mu} \right) = 1.6 \ mW$$

$$\sigma_{DT} = \frac{P_{DT}}{\left(i_O / d_{DO} \right) d_{EI} v_{IN}} = \frac{1.6m}{(10m/1)(51\%)(4)} = 7.8\%$$

Note: P_{DT} is a higher fraction of P_{IN} in DCM than in CCM because P_{IN} scales down faster with i_O than P_{DT} with $\sqrt{i_O}$.

---

5. i_{DS}–v_{DS} Overlap Loss

i_{DS}–v_{DS} overlap power P_{IV} refers to the transitional power transistors consume when switching between on and off states. P_{IV} is essentially the power the *drain–source current* i_{DS} burns across the *drain–source voltage* v_{DS} drops when i_{DS} and v_{DS} transition. This loss hinges on the i_L that i_{DS} carries, the voltage that v_{DS} collapses, and their transition times.

Since i_{DS} scales with *gate–source voltage* v_{GS}, the aim of gate drivers is to transition gate voltages quickly. So they charge and discharge *gate–source* and *–drain oxide capacitances* C_{GS} and C_{GD} with low *pull-down/up N/P-type resistances* R_N and R_P. The resulting transitions should be shorter than dead times, so i_L is practically steady across these events.

5.1. Closing Switch

A. Power

When a switch is open, v_{GS} and i_{DS} are zero and v_{DS} is at a level v_{SW} that other components set. So when closing M_{SW} in Fig. 22, v_{GS} and later i_{DS} climb as R_P charges C_{GS} and C_{GD}. v_{DS} falls when i_{DS} is high enough to sink i_L, C_{GD}'s i_{GD}, and the i_{SW} other switch-node capacitances C_{SW} need to decrease v_{DS}. i_P slews C_{GD} at the v_{GS} (and i_{DS}) needed to sink i_L, i_P, and i_{SW}. So when i_L swamps i_P and i_{SW}, v_{DS} falls after i_{DS} reaches i_L.

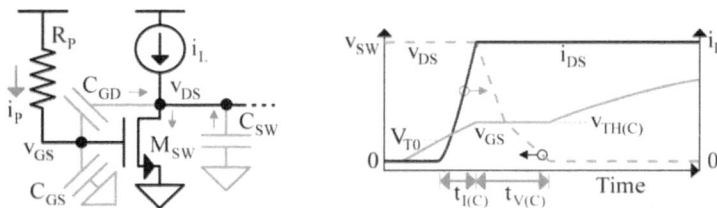

Fig. 22. Closing switch.

Since v_{DS} remains at v_{SW} across the time t_I that i_{DS} needs to reach i_L, v_{SW} burns power P_I across t_I with i_{DS}'s average:

$$P_I = \frac{1}{t_I} \int_0^{t_I} i_{DS} v_{DS} dt$$

$$= \left(\frac{1}{t_I} \int_0^{t_I} i_{DS} dt \right) v_{SW} \approx \left[\frac{1}{t_I} \int_0^{t_I} \left(\frac{i_L}{t_I^2} \right) t^2 dt \right] v_{SW} \approx \left(\frac{i_L}{3} \right) v_{SW} \qquad (36)$$

i_{DS}'s rise is almost quadratic with time t because v_{GS} is close to linear when i_{DS} climbs to i_L and i_{DS} scales with v_{GS}^2 in MOSFET inversion. So across t_I, i_{DS}'s quadratic increase averages about a third of i_L.

i_L also burns power P_V with v_{DS}'s average because i_{DS} is i_L across the time t_V that v_{DS} requires to collapse:

$$P_V = \frac{1}{t_V} \int_0^{t_V} i_{DS} v_{DS} dt \approx i_L \left(\frac{1}{t_V} \int_0^{t_V} v_{DS} dt \right) \approx i_L \left(\frac{v_{SW}}{2} \right). \qquad (37)$$

v_{DS} averages half of v_{SW} because v_{DS} is largely linear. When combined, M_{SW} consumes 33% and 50% of $i_L v_{SW}$ across t_I and t_V and P_{IV} across t_{SW}:

$$P_{IV} = P_I \left(\frac{t_I}{t_{SW}} \right) + P_V \left(\frac{t_V}{t_{SW}} \right) \approx i_L v_{SW} \left(\frac{t_I}{3t_{SW}} + \frac{t_V}{2t_{SW}} \right). \qquad (38)$$

B. Delays

i_{DS} scales with v_{GS} and maxes when v_{GS} reaches the *v_{GS} threshold* v_{TH} needed to sustain i_L, i_P, and i_{SW}. This v_{TH} is usually higher than M_{SW}'s *zero-bias threshold* V_{T0} because i_L is substantial. So M_{SW} inverts in saturation when v_{GS} surpasses V_{T0} because v_{DS} exceeds v_{GS}. When i_L swamps i_P and i_{SW}, the closing v_{TH} is largely the v_{GS} needed to sustain i_L:

$$V_{TH(C)} = V_{T0} + V_{DS(SAT)}\Big|_{i_L + i_P + i_{SW}} \approx V_{T0} + V_{DS(SAT)}\Big|_{i_L} \approx V_{T0} + \sqrt{\frac{2i_L}{K'(W/L)}} , \quad (39)$$

where $v_{DS(SAT)}$ is the *saturation voltage* of M_{SW} in inversion.

i_{DS} does not rise much until M_{SW} inverts. So $t_{I(C)}$ is mostly the time v_{GS} requires to rise from V_{T0} to $v_{TH(C)}$. In other words, t_I is the fraction of the time t_{TH} needed to charge C_{GS} to $v_{TH(C)}$ that excludes the time t_{T0} needed to charge C_{GS} to V_{T0}. Since the gate driver's *power supply* v_{DD} and R_P require RC time t_X to charge C_{GS} and C_{GD} to v_X, t_X and $t_{I(C)}$ are

$$t_X = \tau_{RC} \ln\left(\frac{V_{DD}}{V_{DD} - V_X}\right) \quad (40)$$

and

$$t_{I(C)} \approx t_{TH} - t_{T0} = \tau_{RC(C)} \ln\left(\frac{V_{DD} - V_{T0}}{V_{DD} - V_{TH(C)}}\right), \quad (41)$$

where $\tau_{RC(C)}$ is the time constant of R_P and C_{GS} and C_{GD} in saturation:

$$\tau_{RC(C)} = R_P\left(C_{GS} + C_{GD}\right) = R_P\left[2C_{OL} + \left(\frac{2}{3}\right)C_{CH}\right], \quad (42)$$

and C_{OL} is *overlap capacitance*, C_{CH} is *channel capacitance*, and C_{GS} is C_{OL} plus two–thirds C_{CH} and C_{GD} is C_{OL} in saturated inversion.

v_{DS} falls when i_{DS} sinks i_L, C_{GD}'s i_P, and C_{SW}'s i_{SW}. Although i_{DS} is sensitive to v_{GS} and capable of sinking more current, R_P limits the current that feeds C_{GD}. So the time $t_{V(C)}$ that v_{DS} requires to collapse from v_{SW} is the time C_{GD}'s voltage v_C needs to traverse v_{SW}, which is a slew-rate translation of R_P's current i_P into C_{GD}:

$$t_{V(C)} \approx \left(\frac{\Delta v_C}{i_P}\right)C_{GD}$$

$$\approx \left(\frac{R_P}{V_{DD} - V_{TH(C)}}\right)\left[C_{OL}v_{SW} + \left(\frac{0.5C_{CH}}{2}\right)v_{GS}\right]$$

$$\approx \left(\frac{R_P}{V_{DD} - V_{TH(C)}}\right)\left[C_{OL}v_{SW} + \left(\frac{C_{CH}}{4}\right)v_{TH(C)}\right]. \tag{43}$$

i_P slews C_{GD} at the v_{GS} needed to sustain i_L, i_P, and i_{SW}. So v_{GS} is steady at $v_{TH(C)}$ across t_V and $v_{DD} - v_{TH(C)}$ across R_P sets i_P. As v_{DS} collapses, M_{SW} transitions from saturation to triode, so C_{GD} receives half of C_{CH}. $0.5C_{CH}$'s average $0.25C_{CH}$ therefore traverses the v_{DS} that transitions and keeps M_{SW} in triode, which starts when v_{DS} matches v_{GS} at $v_{TH(C)}$. This is why v_{DS} collapses more quickly at the beginning: because C_{GD}'s saturated C_{OL} is easier to discharge than C_{GD}'s triode $C_{OL} + 0.5C_{CH}$.

Example 9: Determine P_{IV} and σ_{IV} for M_{EG} in the ideal boost from Example 7 in CCM when M_{EG} closes, v_{DD} for M_{EG}'s gate driver is v_O, W is 50 mm, L is 250 nm, L_{OL} is 30 nm, K_N' is 200 µA/V2, C_{OX}'' is 6.9 fF/µm2, V_{TN0} is 400 mV, and R_P is 100 Ω.

Solution:

$d_E = 52\%$, $d_{DO} = d_D = 48\%$, $\Delta i_L = 100$ mA from Example 7

M_{EG} closes when $i_L = i_{L(LO)}$

$$i_{L(LO)} = \frac{i_O}{d_{DO}} - \frac{\Delta i_L}{2} = \frac{250m}{48\%} - \frac{100m}{2} = 470 \text{ mA}$$

$$L_{CH} = L - 2L_{OL} = 250n - 2(30n) = 190 \text{ nm}$$

$$v_{DS(SAT)} \approx \sqrt{\frac{2i_{L(LO)}}{K_N'(W_{CH}/L_{CH})}}$$

$$= \sqrt{\frac{2(470m)}{(200\mu)(50m/190n)}} = 130 \text{ mV}$$

$$V_{TH(C)} \approx V_{TN0} + v_{DS(SAT)} = 400m + 130m = 530 \text{ mV}$$

$$C_{OL} = C_{OX}''W_{CH}L_{OL} = (6.9m)(50m)(30m) = 10 \text{ pF}$$

$$C_{CH} = C_{OX}''W_{CH}L_{CH} = (6.9m)(50m)(190m) = 66 \text{ pF}$$

$$\tau_{RC(C)} = R_P \left[2C_{OL} + \left(\frac{2}{3}\right)C_{CH} \right]$$

$$= (100)\left[2(10p) + \left(\frac{2}{3}\right)(66p) \right] = 6.4 \text{ ns}$$

$$t_{I(C)} \approx \tau_{RC(C)} \ln\left(\frac{V_{DD} - V_{T0}}{V_{DD} - V_{TH(C)}} \right)$$

$$= (6.4p)\ln\left(\frac{4 - 400m}{4 - 530m} \right) = 240 \text{ ps}$$

$$v_{SWO} = v_O + v_{DO} = 4 + 800m = 4.8 \text{ V before } M_{EG} \text{ closes}$$

$$t_{V(C)} \approx \left(\frac{R_P}{V_{DD} - V_{TH(C)}} \right)\left[C_{OL}v_{SWO} + \left(\frac{C_{CH}}{4}\right)V_{TH(C)} \right]$$

$$= \left(\frac{100}{4 - 530m} \right)\left[(10p)(4.8) + \left(\frac{66p}{4}\right)(530m) \right] = 1.6 \text{ ns}$$

$$P_{IV} \approx i_{L(LO)}v_{SWO}\left(\frac{t_{I(C)}}{3t_{SW}} + \frac{t_{V(C)}}{2t_{SW}} \right)$$

$$= (460m)(4.8)\left[\frac{240p}{(3)1\mu} + \frac{1.6n}{(2)1\mu} \right] = 2.0 \text{ mW}$$

$$\sigma_{IV} = \frac{P_{IV}}{\left(i_O/d_{DO}\right)d_{EI}V_{IN}} = \frac{2.0m}{(250m/48\%)(1)(2)} = 0.2\%$$

Note: This P_{IV} excludes the power consumed when M_{EG} opens and other switches open and close.

5.2. Opening Switch

A. Power

After M_{SW} closes, v_{GS} is high, i_{DS} carries i_L, and v_{DS} nears zero. To open M_{SW}, R_N in Fig. 23 collapses v_{GS}. i_{DS}, however, does not fall until v_{GS} drops below the v_{TH} needed to sink i_L minus C_{GD}'s i_N, and C_{SW}'s i_{SW}. Even then, i_N slews C_{GD} at the v_{GS} (and i_{DS}) needed to sustain this i_{DS}.

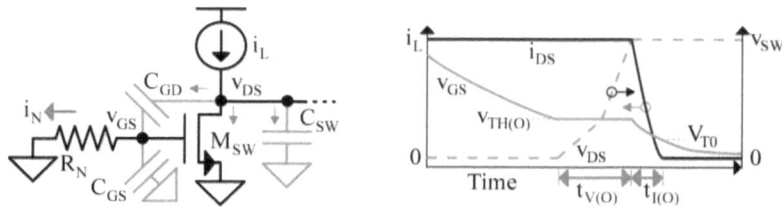

Fig. 23. Opening switch.

So v_{DS} rises across the $t_{V(O)}$ that i_N needs to charge C_{GD} and raise v_{DS} to a level v_{SW} that other components set. P_V is the power v_{DS}'s average 50%v_{SW} burns across $t_{V(O)}$ with i_{DS} at i_L minus i_N and i_{SW}. When i_L swamps i_N and i_{SW}, v_{TH} is largely the v_{GS} needed to sustain i_L:

$$v_{TH(O)} = V_{T0} + V_{DS(SAT)}\Big|_{i_L - i_N - i_{SW}} \approx V_{T0} + V_{DS(SAT)}\Big|_{i_L} \approx V_{T0} + \sqrt{\frac{2i_L}{K'(W/L)}} . \quad (44)$$

Since v_{DS} stops rising at v_{SW}, the effect of i_N on C_{GD} after $t_{V(O)}$ is to decrease v_{GS}. For this to happen, part of i_N must also discharge C_{GS}. The end result is that i_{DS} drops across the $t_{I(O)}$ that i_N needs to reduce v_{GS} from $v_{TH(O)}$ to V_{T0}. This reduction in v_{GS} is so much lower than v_{GS}'s total swing that v_{GS} is close to linear across $t_{I(O)}$. So P_I is roughly the power i_{DS}'s quadratic average 33%i_L burns across $t_{I(O)}$ with v_{DS} at v_{SW}.

B. Delays

Since v_{GS} is $v_{TH(O)}$ across $t_{V(O)}$, i_N is an R_N translation of $v_{TH(O)}$. So $v_{TH(O)}/R_N$ slews C_{GD}'s C_{OL} across v_{SW} and C_{GD}'s channel average 0.25C_{CH} across the v_{DS} that transitions M_{SW} into triode, which starts when v_{DS} matches v_{GS}'s $v_{TH(O)}$. The resulting $t_{V(O)}$ is

$$t_{V(O)} \approx \left(\frac{\Delta v_C}{i_D}\right)C_{GD} \approx \left(\frac{R_N}{v_{TH(O)}}\right)\left[C_{OL}v_{SW} + \left(\frac{C_{CH}}{4}\right)v_{TH(O)}\right]. \quad (45)$$

Once v_{DS} reaches v_{SW}, R_N's current i_N reduces v_{GS}. So i_{DS} falls across the $t_{I(O)}$ that R_N needs to discharge C_{GS} (and C_{GD}) from $v_{TH(O)}$ to V_{T0}. In other words, $t_{I(O)}$ is the part of the time t_{T0} needed to discharge C_{GS} to V_{T0} that excludes the time t_{TH} needed to discharge C_{GS} to $v_{TH(O)}$. C_{GS} therefore discharges $v_{DD} - v_{TH(O)}$ across t_{TH}, $v_{DD} - V_{T0}$ across t_{T0}, and $v_{TH(O)} - V_{T0}$ across their difference $t_{T0} - t_{TH}$ or $t_{I(O)}$:

$$t_{I(O)} \approx t_{T0} - t_{TH} = \tau_{RC(O)} \ln\left[\frac{v_{DD} - \left(v_{DD} - v_{TH(O)}\right)}{v_{DD} - \left(v_{DD} - V_{T0}\right)}\right] = \tau_{RC(O)} \ln\left(\frac{v_{TH(O)}}{V_{T0}}\right), \quad (46)$$

where $\tau_{RC(O)}$ is the time constant of R_N and C_{GS} and C_{GD} in saturation:

$$\tau_{RC(O)} = R_N\left(C_{GS} + C_{GD}\right) = R_N\left[2C_{OL} + \left(\frac{2}{3}\right)C_{CH}\right]. \quad (47)$$

Note that R_P pre-charges C_{GS} to v_{DD} before R_N collapses C_{GS}'s v_{GS}.

Example 10: Determine P_{IV} and σ_{IV} for M_{EG} in the boost from Examples 7 and 9 in CCM when M_{EG} opens and R_N is 20 Ω.

Solution:

$d_E = 52\%$, $d_{DO} = d_D = 48\%$, $\Delta i_L = 100$ mA from Example 7

$L_{CH} = 190$ nm, $C_{OL} = 10$ pF, $C_{CH} = 66$ pF from Example 9

M_{EG} opens when $i_L = i_{L(HI)}$

$$i_{L(HI)} = \frac{i_O}{d_{DO}} + \frac{\Delta i_L}{2} = \frac{250m}{48\%} + \frac{100m}{2} = 570 \ mA$$

$$V_{DS(SAT)} \approx \sqrt{\frac{2i_{L(HI)}}{K_N'(W_{CH}/L_{CH})}}$$

$$= \sqrt{\frac{2(570m)}{(200\mu)(50m/190n)}} = 150 \ mV$$

$v_{TH(O)} \approx V_{TN0} + v_{DS(SAT)} = 400m + 150m = 550 \text{ mV}$

$v_{SWO} = v_O + v_{DO} = 4 + 800m = 4.8 \text{ V}$ after M_{EG} opens

$$t_{V(O)} \approx \left(\frac{R_N}{v_{TH(O)}} \right) \left[C_{OL} v_{SWO} + \left(\frac{C_{CH}}{4} \right) v_{TH(O)} \right]$$

$$= \left(\frac{20}{550m} \right) \left[(10p)(4.8) + \left(\frac{66p}{4} \right)(550m) \right] = 2.1 \text{ ns}$$

$$\tau_{RC(O)} = R_N \left[2C_{OL} + \left(\frac{2}{3} \right) C_{CH} \right]$$

$$= (20) \left[2(10p) + \left(\frac{2}{3} \right)(66p) \right] = 1.3 \text{ ns}$$

$$t_{I(O)} \approx \tau_{RC(O)} \ln \left(\frac{v_{TH(O)}}{V_{TN0}} \right) = (1.3n) \ln \left(\frac{550m}{400m} \right) = 410 \text{ ps}$$

$$P_{IV} \approx i_{L(HI)} v_{SWO} \left(\frac{t_{I(O)}}{3t_{SW}} + \frac{t_{V(O)}}{2t_{SW}} \right)$$

$$= (570m)(4.8) \left[\frac{410p}{(3)1\mu} + \frac{2.1n}{(2)1\mu} \right] = 3.2 \text{ mW}$$

$$\sigma_{IV} = \frac{P_{IV}}{\left(i_O / d_{DO} \right) d_{EI} v_{IN}} = \frac{3.2m}{(250m/48\%)(1)(2)} = 0.3\%$$

Note: R_N is lower than R_P to balance response times (because $v_{TH(O)}$ is more limiting than $v_{DD} - v_{TH(C)}$ in t_V and v_{GS} reaches V_{TN0} near the end of R_N's exponential response, where the response is slower). M_{EG} still consumes more power opening than closing because i_{DS} is higher at $i_{L(HI)}$.

5.3. Reverse Recovery

A. Power

Forward-biased in-transit charge across PN junctions reverses direction when diodes reverse-bias. In switched inductors, dead-time diodes carry this *reverse-recovery charge* q_{RR} when conducting i_L. q_{RR} is the charge in the junction that i_L feeds and *forward transit time* τ_F across the junction sets to $i_L\tau_F$.

The challenge with q_{RR} is that a switch must recover it. In Fig. 24, for example, *dead-time diode* D_{DT} conducts i_L when M_{SW} is open. So for v_{DS} to fall when M_{SW} closes, i_{DS} must first rise to a peak $i_{DS(RR)}$ that sinks i_L, i_{GD}, and q_{RR} held in D_{DT}. And a higher i_{DS} dissipates more P_{IV}.

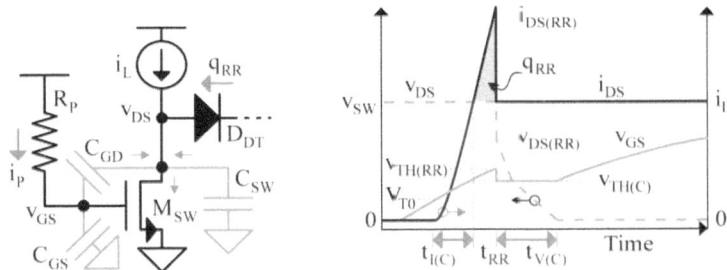

Fig. 24. Closing switch with reverse recovery.

As M_{SW} closes, i_{DS} climbs with v_{GS} mostly after v_{GS} overcomes V_{T0}. When i_L swamps i_P and i_{SW}, i_{DS} reaches i_L after $t_{I(C)}$ when v_{GS} reaches $v_{TH(C)}$. Approximating i_{DS}'s rise past i_L to be linear with the slope $\partial i_{DS}/\partial t$ that i_{DS}'s quadratic climb $(i_L/t_{I(C)}^2)t^2$ reaches at $t_{I(C)}$, i_{DS} requires another t_{RR} to reach a level that can sink q_{RR}:

$$q_{RR} = \int_0^{t_{RR}} i_{DS}dt \approx \int_0^{t_{RR}} \frac{\partial i_{DS}}{\partial t}\bigg|_{t_{I(C)}} tdt \approx \int_0^{t_{RR}} \left(\frac{2i_L}{t_{I(C)}}\right)tdt = \left(\frac{i_L}{t_{I(C)}}\right)t_{RR}^2 = i_L\tau_F . \quad (48)$$

This means that t_{RR} is roughly a squared-root translation of $t_{I(C)}$ and τ_F:

$$t_{RR} \approx \sqrt{t_{I(C)}\tau_F} , \quad (49)$$

and $i_{DS(RR)}$ is a corresponding t_{RR} extension of i_L and i_P when i_{SW} is low:

$$i_{DS(RR)} \approx i_L + i_P + \left(\frac{2i_L}{t_{I(C)}} \right) t_{RR} = i_P + i_L \left(1 + 2\sqrt{\frac{\tau_F}{t_{I(C)}}} \right). \tag{50}$$

The v_{GS} that M_{SW} requires to sink this $i_{DS(RR)}$ is $v_{TH(RR)}$:

$$v_{TH(RR)} = V_{T0} + V_{DS(SAT)}\Big|_{i_{DS(RR)}} = V_{T0} + \sqrt{\frac{2i_{DS(RR)}}{K'(W/L)}}. \tag{51}$$

Since v_{DS} is steady at v_{SW} across $t_{I(C)}$ and t_{RR}, P_I' is the power $i_{DS(RR)}$'s average 33%$i_{DS(RR)}$ burns with v_{SW} across $t_{I(C)}$ and t_{RR}.

After i_{DS} recovers q_{RR}, $i_{DS(RR)}$ sinks more than i_L supplies, so i_{DS} discharges C_{GD} and C_{SW}. The i_{GD} that $i_{DS(RR)}$ and i_L avail is so much greater than i_P that i_{GD} discharges C_{GS}. C_{GD}, C_{SW}, and C_{GS} discharge this way until i_{DS} falls to i_L plus i_P and i_{SW}, which happens when v_{GS} reaches $v_{TH(C)}$. Since i_P is much lower than i_{GD}, C_{GS} supplies the charge Δq_{GS} that largely discharges C_{GD} across Δv_{DG} or $\Delta v_{DS} - \Delta v_{GS}$:

$$\begin{aligned} \Delta q_{GD} &= C_{GD}\Delta v_{DG} = C_{GD}\left(\Delta v_{DS} - \Delta v_{GS} \right) \\ &= C_{GD}\left(v_{SW} - v_{DS(RR)} - \Delta v_{GS} \right) \\ &\approx \Delta q_{GS} = C_{GS}\Delta v_{GS} = C_{GS}\left(v_{TH(RR)} - v_{TH(C)} \right) = C_{GS}\Delta v_{TH} \end{aligned} \tag{52}$$

where C_{GS} discharges across Δv_{GS} or Δv_{TH} from $v_{TH(RR)}$ to $v_{TH(C)}$ and v_{DS} falls across Δv_{DS} from v_{SW} to $v_{DS(RR)}$. Solving this reveals v_{DS} falls to

$$v_{DS(RR)} \approx v_{SW} - \left(\frac{C_{GS}}{C_{GD}} + 1 \right) \Delta v_{TH}. \tag{53}$$

This transition to $v_{DS(RR)}$ is quick because i_{GD} is substantial.

After i_{DS} falls to i_L plus i_P and i_{SW}, i_P discharges C_{GD} across the $t_{V(C)}'$ that collapses $v_{DS(RR)}$. $t_{V(C)}'$ is shorter than $t_{V(C)}$ because v_{DS} collapses $v_{DS(RR)}$, which is lower than v_{SW}:

$$t_{V(C)}' \approx \left(\frac{R_P}{V_{DD} - V_{TH(C)}} \right) \left[C_{OL} V_{DS(RR)} + \left(\frac{C_{CE}}{4} \right) V_{TH(C)} \right]. \qquad (54)$$

Since i_{DS} is steady at i_L across $t_{V(C)}'$, P_V' is the power i_L burns with $v_{DS(RR)}$'s average 50%$v_{DS(RR)}$. And P_{IV}' is a t_{SW} fraction of P_I' and P_V':

$$P_{IV}' = P_I' \left(\frac{t_{I(C)} + t_{RR}}{t_{SW}} \right) + P_V' \left(\frac{t_{V(C)}'}{t_{SW}} \right)$$

$$\approx \left(\frac{i_{DS(RR)}}{3} \right) V_{SW} \left(\frac{t_{I(C)} + t_{RR}}{t_{SW}} \right) + i_L \left(\frac{V_{DS(R3)}}{2} \right) \left(\frac{t_{V(C)}'}{t_{SW}} \right) \qquad (55)$$

Interestingly, q_{RR} not only raises the i_{DS} that consumes P_I' (to $i_{DS(RR)}$) and extends the time P_I' burns (by t_{RR}) but also reduces the v_{DS} (to $v_{DS(RR)}$) that burns P_V'. The rise in i_{DS}, however, normally raises P_I' more than the fall in $v_{DS(RR)}$ reduces P_V'. So reducing q_{RR} usually saves power.

Example 11: Determine P_{IV} and σ_{IV} for M_{EG} in the ideal boost from Examples 7 and 9 in CCM when M_{EG} closes and D_{DO}'s τ_F is 5 ns.

Solution:

$$d_E = 52\%, d_{DO} = d_D = 48\%, \Delta i_L = 100 \text{ mA from Example 7}$$

$$i_{L(LO)} = 470 \text{ mA}, v_{TH(C)} = 530 \text{ mV}, v_{SWO} = 4.8 \text{ V},$$

$$L_{CH} = 190 \text{ nm}, C_{OL} = 10 \text{ pF}, C_{CH} = 66 \text{ pF},$$

$$R_P = 100 \text{ }\Omega, \tau_{RC} = 6.4 \text{ ns}, t_{I(C)} \approx 240 \text{ ps from Example 9}$$

$$t_{RR} \approx \sqrt{t_{I(C)} \tau_F} = \sqrt{(240p)(5n)} = 1.1 \text{ ns}$$

$$i_P \approx \frac{V_{DD} - V_{TH(C)}}{R_P} = \frac{4 - 530m}{100} = 35 \text{ mA}$$

$$i_{DS(RR)} \approx i_{L(LO)} + i_P + \left(\frac{2i_{L(LO)}}{t_{I(C)}} \right) t_{RR}$$

$$= (470\text{m})\left[1 + 2\left(\frac{1.1\text{n}}{240\text{p}}\right)\right] + 35\text{m} = 4.8 \text{ A}$$

$$V_{TH(RR)} = V_{TN0} + \sqrt{\frac{2i_{DS(RR)}}{K_N'(W_{CH}/L_{CH})}}$$

$$= 400\text{m} + \sqrt{\frac{2(4.8)}{(200\mu)(50\text{m}/190\text{n})}} = 830 \text{ mV}$$

$$V_{DS(RR)} = v_{SWO} - \left(\frac{C_{OL} + (2/3)C_{CH}}{C_{OL}} + 1\right)\left(V_{TH(RR)} - V_{TH(C)}\right)$$

$$= 4.8 - \left[\frac{10\text{p} + (2/3)(66\text{p})}{10\text{p}} + 1\right](830\text{m} - 530\text{m})$$

$$= 2.9 \text{ V}$$

$$t_{V(C)}' \approx \left(\frac{1}{i_P}\right)\left[C_{OL}V_{DS(RR)} + \left(\frac{C_{CH}}{4}\right)V_{TH(C)}\right]$$

$$= \left(\frac{1}{35\text{m}}\right)\left[(10\text{p})(2.9) + \left(\frac{66\text{p}}{4}\right)(530\text{m})\right] = 1.1 \text{ ns}$$

$$P_{IV} \approx \left(\frac{i_{DS(RR)}}{3}\right)v_{SWO}\left(\frac{t_{I(C)} + t_{RR}}{t_{SW}}\right) + i_L\left(\frac{V_{DS(RR)}}{2}\right)\left(\frac{t_{V(C)}'}{t_{SW}}\right)$$

$$= \left(\frac{4.8}{3}\right)(4.8)\left(\frac{240\text{p} + 1.1\text{n}}{1\mu}\right) + (470\text{m})\left(\frac{2.9}{2}\right)\left(\frac{1.1\text{n}}{1\mu}\right)$$

$$= 11 \text{ mW}$$

$$\sigma_{IV} = \frac{P_{IV}}{(i_O/d_{DO})d_{EI}v_{IN}} = \frac{11\text{m}}{(250\text{m}/48\%)(1)(2)} = 1.1\%$$

Note: This P_{IV} is higher than in Example 9 because $i_{DS(RR)}$ is higher than $i_{L(LO)}$ and $t_{I(C)} + t_{RR}$ is longer than $t_{I(C)}$ to a greater extent than $v_{DS(RR)}$ is lower than v_{SWO}.

B. MOS Diodes

Reverse-recovery charge is greatest when body diodes conduct i_L, which happens when switches are off. M_{DG} and M_{DO} open in Fig. 25, for example, when the controller grounds M_{DG}'s gate and connects M_{DO}'s gate to v_O. This way, dead-time i_L discharges v_{SWI}'s C_{SWI} and charges v_{SWO}'s C_{SWO} until M_{DG}'s and M_{DO}'s body diodes conduct i_L.

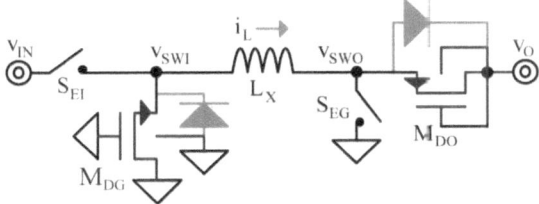

Fig. 25. Switched inductor during dead time.

When these body diodes conduct, v_{SWI} is a diode voltage below ground and v_{SWO} is a diode voltage above v_O. Interestingly, M_{DG}'s v_{GS} and M_{DO}'s v_{SG} are positive under these conditions. Although sub-threshold current is not always negligible, M_{DG} and M_{DO} remain largely off when their threshold voltages match or surpass a PN diode voltage.

M_{DG} and M_{DO} start conducting some of i_L when threshold voltages are lower. So the current that feeds q_{RR} drops as V_{T0}'s fall below 600–700 mV. This diminishes the effect of q_{RR} on P_{IV}. The effect is minimal when the v_{GS} and v_{SG} that M_{DG} and M_{DO} need to sustain i_L are less than 600–700 mV, which can happen when V_{T0}'s are 200–300 mV.

Such low V_{T0}'s are uncommon because large low-V_{T0} switches leak considerable current with zero v_{GS}. Although technology and application vary, 400–600-mV V_{T0}'s are more common. With these V_{T0}'s, the effect of q_{RR} on P_{IV} is still present, but not as severe.

C. Schottky Diodes

Letting body diodes conduct poses another challenge: *substrate noise*. This is because MOSFETs on the substrate conduct current through the substrate. And MOSFETs in wells over the substrate activate vertical *bipolar-junction transistors* (BJTs) that inject current into the substrate. The problem is that switching events produce, inject, and propagate noise energy across the substrate, coupling into many circuits along the way.

Connecting *Schottky diodes* in parallel with M_{DG}'s and M_{DO}'s body diodes like Fig. 26 shows reduces this problem. This helps because they drop lower voltages than typical PN diodes without a junction that traps in-transit charge. So they steer current away from body diodes without injecting noise and without imposing reverse-recovery charge.

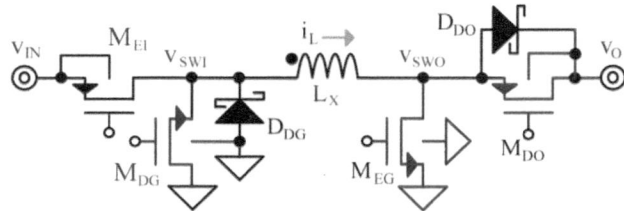

Fig. 26. Switched inductor with Schottky diodes.

Good Schottky diodes, however, are not always available on-chip. And off-chip diodes require board space and money. Still, the benefits of lower noise and lower power can outweigh the added volume and cost.

5.4. Soft Switching

P_{IV} hinges on the currents and voltages that power switches carry and collapse. *Soft switching* refers to transitions that carry low currents or collapse low voltages. Although *zero-current switching* (ZCS) and *zero-voltage switching* (ZVS) are extreme cases, engineers often use ZVS and ZCS to refer to "soft" events. Either way, the net result is low P_{IV}.

A. Zero-Voltage Switching

Collapsing v_{IN} or v_O is not ZVS. Energize switches fall into this category. M_{EI}, for example, collapses v_{IN} plus v_{DG} because M_{EI} connects v_{IN} to v_{SWI} and D_{DG} conducts i_L before and after M_{EI} closes. M_{EG} similarly collapses

v_O and v_{DO} because D_{DO} conducts i_L into v_O before and after M_{EG} closes. These are *hard-switching* examples.

Collapsing diode voltages like drain switches do is "soft". M_{DG}, for one, collapses D_{DG}'s v_{DG} when M_{DG} grounds v_{SWI} a t_{DT} before and after t_E in Fig. 27. And M_{DO} collapses D_{DO}'s v_{DO} when M_{EO} connects v_{SWO} to v_O at those same times. These transitions are practically ZVS when v_{DG} and v_{DO} are small fractions of v_{IN} and v_O, so P_{IV} is usually low.

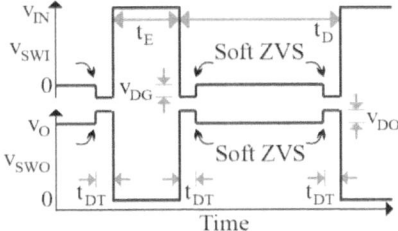

Fig. 27. Soft zero-voltage switching events.

B. Zero-Current Switching

ZCS happens in DCM when i_L is zero. Drain switches, for example, open with ZCS because i_L reaches zero when t_D ends in Fig. 28. Similarly, energize switches close with ZCS because i_L is zero when t_E starts. So P_{IV} when t_D ends and t_E starts is negligibly low.

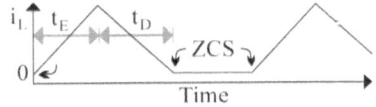

Fig. 28. Zero-current switching events.

5.5. CMOS Expressions

NFETs are *active high*, which means they close when their gate voltages are high. Discussions, graphs, and expressions to this point reflect this context. In this light, v_{GS}, v_{DS}, and V_{T0} are positive values.

PFETs behave like NFETs, except they are *active low*, which is to say they close with low gate voltages. So pull-down resistors close them and pull-up resistors open them. CMOS stands for *complementary MOS transistors* for this reason: because NFETs and PFETs are available.

Not coincidentally, NFET expressions also apply to PFETs. Except, v_{GS}, v_{DS}, and V_{T0} for PFETs are negative, which is not always intuitive. Replacing v_{GS}, v_{DS}, and V_{T0} with v_{SG}, v_{SD}, and $|V_{T0}|$ or $|V_{TP0}|$ is more insightful because PFETs collapse v_{SD} when v_{SG} overcomes $|V_{T0}|$.

Example 12: Determine P_{IV} and σ_{IV} for M_{EI} in the ideal buck in CCM when M_{EI} closes, v_{DD} for M_{EI}'s gate driver is v_{IN}, v_{IN} is 4 V, v_O is 2 V, v_{DG} is 800 mV, i_O is 250 mA, L_X is 10 μH, t_{SW} is 1 μs, t_{DT} is 50 ns, W is 50 mm, L is 250 nm, L_{OL} is 30 nm, K_P' is 40 μA/V^2, C_{OX}'' is 6.9 fF/μm^2, V_{TP0} is −400 mV, and R_N is 50 Ω.

Solution:

$$d_{EI} = d_E = \frac{v_O}{v_{IN}} + \left(\frac{v_{DG}}{v_{IN}} \right)\left(\frac{2t_{DT}}{t_C} \right)$$

$$= \frac{2}{4} + \left(\frac{800m}{4} \right)\left[\frac{2(50n)}{1\mu} \right] = 52\%$$

$$\therefore \quad d_{DO} = d_D = 1 - d_E = 1 - 52\% = 48\%$$

$$\Delta i_L = \left(\frac{v_E}{L_X} \right) d_E t_{SW} = \left(\frac{v_{IN} - v_O}{L_X} \right) d_E t_{SW}$$

$$\approx \left(\frac{4-2}{10\mu} \right)(52\%)(1\mu) = 100 \ mA$$

$$i_{L(LO)} = \frac{i_O}{d_{DO}} - \frac{\Delta i_L}{2} = \frac{250m}{100\%} - \frac{100m}{2} = 200 \ mA$$

W, L, L_{OL}, and C_{OX}'' match M_{EG}'s

$$\therefore \quad L_{CH} = 190 \ nm, C_{OL} = 10 \ pF, C_{CH} = 66 \ pF$$

from Example 9

M_{EI} closes when $i_L = i_{L(LO)}$

$$v_{SD(SAT)} \approx \sqrt{\frac{2i_{L(LO)}}{K_P'(W_{CH}/L_{CH})}}$$

$$= \sqrt{\frac{2(200m)}{(40\mu)(50m/190n)}} = 200 \text{ mV}$$

$v_{TH(C)} \approx |V_{TP0}| + v_{SD(SAT)} = 400m + 200m = 600 \text{ mV}$

$$\tau_{RC(C)} = R_N\left[2C_{OL} + \left(\frac{2}{3}\right)C_{CH}\right]$$

$$= (50)\left[2(10p) + \left(\frac{2}{3}\right)(66p)\right] = 3.2 \text{ ns}$$

$$t_{I(C)} \approx \tau_{RC(C)} \ln\left(\frac{V_{DD} - |V_{TP0}|}{V_{DD} - V_{TH(C)}}\right)$$

$$= (3.2n)\ln\left(\frac{2 - 400m}{2 - 600m}\right) = 180 \text{ ps}$$

$v_{SWI} = -v_{DG} = -800 \text{ mV}$ before M_{EI} closes

$v_{SWI} \approx v_{IN} = 4 \text{ V}$ after M_{EI} closes

$\therefore \quad v_{SD}$ swings $v_{IN} - (-v_{DG}) = v_{IN} + v_{DG} = 4.8 \text{ V}$

$$t_{V(C)} \approx \left(\frac{R_N}{V_{DD} - V_{TH(C)}}\right)\left[C_{OL}\left(v_{IN} + v_{DG}\right) + \left(\frac{C_{CH}}{4}\right)v_{TH(C)}\right]$$

$$= \left(\frac{50}{2 - 600m}\right)\left[(10p)(4.8) + \left(\frac{66p}{4}\right)(600m)\right] = 850 \text{ ps}$$

$$P_{IV} \approx i_{L(LO)}\left(v_{IN} + v_{DG}\right)\left(\frac{t_{I(C)}}{3t_{SW}} + \frac{t_{V(C)}}{2t_{SW}}\right)$$

$$= (200m)(4.8)\left[\frac{180p}{(3)1\mu} + \frac{850p}{(2)1\mu}\right] = 470 \text{ }\mu W$$

$$\sigma_{IV} = \frac{P_{IV}}{\left(i_O/d_{DO}\right)d_{EI}v_{IN}} = \frac{470\mu}{(250m/1)(52\%)(4)} = 0.1\%$$

Note: This P_{IV} excludes the power consumed when M_{EI} opens and other switches open and close.

---

6. Gate-Driver Loss

6.1. Gate Driver

Typical gate drivers use pull-up PFETs to charge gates and pull-down NFETs to discharge gates. Since NFETs are active high and PFETs are active low, a high input v_I in Fig. 29 activates M_N and shuts M_P, so M_N grounds the output v_O. A low v_I does the opposite: shuts M_N and activates M_P, so M_P pulls v_O to the power supply v_{DD}.

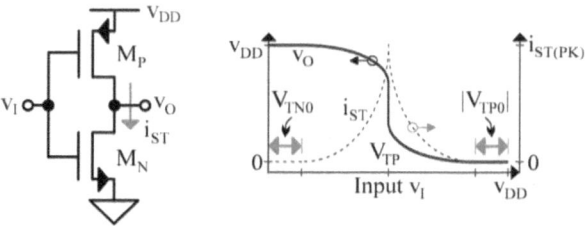

Fig. 29. Inverting gate driver.

As v_I climbs from zero, M_N starts to conduct when v_I surpasses V_{TN0}. v_O transitions low when M_N pulls as much current as M_P can supply. The *shoot-through current* i_{ST} that M_N and M_P conduct maxes at this point. The *trip point* V_{TP} of the driver is the v_I that balances M_N's and M_P's strengths this way. So i_{ST} maxes when v_I reaches V_{TP} and v_O halves v_{DD}:

$$i_{ST(MAX)} = i_N \Big|_{v_{DS} = \frac{v_{DD}}{2}}^{v_{GS} = V_{TP}}$$

$$= \left(\frac{W_N}{L_N}\right)\left(\frac{K_N{'}}{2}\right)(V_{TP} - V_{TN0})^2 \left[1 + \left(\frac{v_{DD}}{2}\right)\lambda_N\right]$$

$$= i_P \Big|_{v_{SD} = \frac{v_{DD}}{2}}^{v_{SG} = V_{DD} - V_{TP}}$$

$$= \left(\frac{W_P}{L_P}\right)\left(\frac{K_P{}'}{2}\right)\left(V_{DD} - V_{TP} - |V_{TP0}|\right)^2\left[1 + \left(\frac{V_{DD}}{2}\right)\lambda_P\right], \quad (56)$$

where v_I at V_{TP} and v_O at $0.5v_{DD}$ invert and saturate M_N and M_P.

A few points are worth noting. V_{TP} is $0.5v_{DD}$ when V_{T0}'s and *channel-length modulation parameters* λ's match and M_P's W/L is greater than M_N's by the same amount that M_N's *transconductance parameter* K' is greater than M_P's. v_O can rail to zero and v_{DD}. Static power when v_I is within a V_{T0} of the supplies is nearly zero because i_{ST} is almost zero. This is why this circuit is so attractive as a *digital inverter*.

The *gate capacitances* C_G that load the driver need M_P's i_P to charge and M_N's i_N to discharge. This means that i_N wastes power when charging and i_P wastes power when discharging. But if M_N's v_{DS} is low when C_G charges and M_P's v_{SD} is low when C_G discharges, their triode currents waste less power. This happens when v_O's transition is slower than v_I's, which results when M_P and M_N charge and discharge C_G slowly.

Across the t_I and t_V that switches carry i_L, the driver's v_O is V_{T0} to v_{TH} above ground or below v_{DD}. Since M_N's v_{GS} is v_{DD} and $v_{DS(SAT)}$ is $v_{DD} - V_{TN0}$, M_N's v_{DS} is lower than $v_{DS(SAT)}$ across t_I and t_V. M_P's v_{SD} is similarly lower than $v_{SD(SAT)}$'s $v_{DD} - |V_{TP0}|$. So M_N and M_P are in triode across t_I and t_V, when v_{DS} is largely v_{TH} or $v_{DD} - v_{TH}$, which means their triode resistances set the R_N and R_P that open and close power transistors:

$$R_{N/P} = \frac{v_{DS}}{i_{TRI}} = \frac{1}{\left(W_{CH}/L_{CH}\right)K'\left(v_{DD} - V_{T0} - 0.5v_{DS}\right)}. \quad (57)$$

Example 13: Determine the W's for the gate driver that closes and opens M_{EG} in the ideal boost from Examples 7, 9, and 10 when v_{DD} is v_O, L's are 250 nm, L_{OL} is 30 nm, V_{TN0} is 400 mV, V_{TP0} is -400 mV, K_N' is 200 µA/V^2, and K_P' is 40 µA/V^2.

Solution:

$$v_{DD} = v_O = 4 \text{ V}, R_P = 100 \text{ } \Omega, v_{TH(C)} = 530 \text{ mV}, R_N = 20 \text{ } \Omega,$$

$$v_{TH(O)} = 550 \text{ mV from Examples 9 and 10}$$

$$L_{CH} = L - 2L_{OL} = 250n - 2(30n) = 190 \text{ nm}$$

M_P's $v_{SD} = v_{DD} - v_{TH(C)}$

$$W_P = \frac{L_{CH}}{R_U K_P' \left[v_{DD} - |V_{TP0}| - 0.5\left(v_{DD} - v_{TH(C)} \right) \right]}$$

$$= \frac{190n}{(100)(40\mu)\left[4 - 400m - 0.5(4 - 530m) \right]} = 26 \text{ } \mu m$$

M_N's $v_{DS} = v_{TH(O)}$

$$W_N = \frac{L_{CH}}{R_D K_N' \left(v_{DD} - V_{TN0} - 0.5 v_{TH(O)} \right)}$$

$$= \frac{190n}{(20)(200\mu)\left[4 - 400m - 0.5(550m) \right]} = 14 \text{ } \mu m$$

6.2. Closing Switch

When closing M_{SW} in Fig. 30, M_P supplies the *gate current* i_G that C_{GB}, C_{GS}, and C_{GD} need and the i_{ST} that M_N leaks. i_{ST} is low and short-lived (by design) because M_{SW}'s low v_{GS} suppresses M_N's v_{DS} as v_I's fall collapses M_N's v_{GS}. i_G is much greater because M_P's v_{SG} and v_{SD} are high across the time M_N leaks i_{ST}. So the *driver current* i_D that v_{DD} supplies is mostly the charge q_G that C_{GB}, C_{GS}, and C_{GD} need to close M_{SW}.

For v_I to fall in the first place, the pre-driver must sink the charge q_{GI} that M_N and M_P's gate capacitances C_{GI} store when v_I is at v_{DD}. This means that the pre-driver drains and burns C_{GI}'s energy. The *drive power* P_D that v_{DD} supplies therefore reduces to the power M_{SW} needs to close:

$$P_{D(C)} = P_G + P_{ST(C)} \approx P_G, \tag{58}$$

where *gate-charge power* P_G is the power v_{DD} supplies with q_G:

$$P_G = v_{DD} i_{G(AVG)} = v_{DD} \left(\frac{q_G}{t_{SW}} \right), \tag{59}$$

and $P_{ST(C)}$ is the power M_N leaks with the i_{LK} that v_{DD} supplies across t_{ST}:

$$P_{ST} = v_{DD} i_{LK(AVG)} = v_{DD} \left(\frac{1}{t_{SW}} \int_0^{t_{LK}} i_{LK} \, dt \right). \tag{60}$$

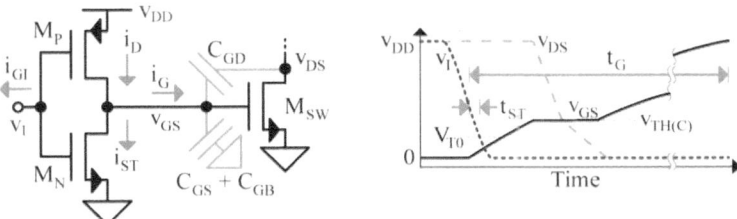

Fig. 30. Gate driver closing switch.

Components in C_{GB} and C_{GS} change as v_{GS} traverses v_{DD}. This is because v_{GS}'s rise inverts M_{SW} into saturation (because v_{DS} is high) and v_{DS}'s fall then collapses M_{SW} into triode. So C_{GB}'s C_{CH} charges to V_{T0}, C_{GS}'s C_{OL} charges to V_{T0} when off, C_{GS}'s C_{OL} and two–thirds C_{CH} charge from V_{T0} to $v_{TH(C)}$ in saturation, and C_{GS}'s C_{OL} and half C_{CH} charge from $v_{TH(C)}$ to v_{DD} in triode. The charge q_{GBS} that C_{GB} and C_{GS} draw is

$$q_{GBS} \approx C_{CH} V_{T0} + C_{OL} V_{DD}$$
$$+ \left(\frac{2}{3} \right) C_{CH} \left(v_{TH(C)} - V_{T0} \right) + \left(\frac{1}{2} \right) C_{CH} \left(v_{DD} - v_{TH(C)} \right). \tag{61}$$

i_G not only raises C_{GD}'s v_{GS} across v_{DD} but also collapses v_{DS} across v_{SW}, which means C_{GD} charges across v_{DD} and v_{SW}. C_{GD}'s C_{OL} charges with v_{GS} to v_{DD} and with v_{DS} across v_{SW}. Since C_{GD} begins to receive half C_{CH} when v_{DS} falls below v_{GS}'s $v_{TH(C)}$, $0.5C_{CH}$'s average $0.25C_{CH}$ charges across $v_{TH(C)}$. After v_{DS} falls, C_{GD}'s $0.5C_{CH}$ charges with v_{GS} from $v_{TH(C)}$ to v_{DD} in triode:

$$q_{GD} \approx C_{OL}\left(v_{DD}+v_{SW}\right)+\left(\frac{1}{4}\right)C_{CH}v_{TH(C)}+\left(\frac{1}{2}\right)C_{CH}\left(v_{DD}-v_{TH(C)}\right). \quad (62)$$

In all, i_G delivers q_{GBS} and q_{GD}. Since v_{TH} is the v_{GS} needed to sustain i_L, which is $v_{DS(SAT)}$ over V_{T0}, q_G reduces to

$$q_G = q_{GBS}+q_{GD}$$

$$= C_{OL}\left(2v_{DD}+v_{SW}\right)+C_{CH}\left(v_{DD}+\frac{V_{T0}}{4}-\frac{v_{DS(SAT)}}{12}\right). \quad (63)$$

$$\approx C_{OL}\left(2v_{DD}+v_{SW}\right)+C_{CH}\left(v_{DD}+\frac{V_{T0}}{4}\right)$$

Since $v_{DS(SAT)}$ is usually lower than V_{T0}, a twelfth is negligibly lower.

6.3. Opening Switch

When opening M_{SW} in Fig. 31, M_N sinks the i_G that drains C_{GB}, C_{GS}, and C_{GD} and the i_{ST} that M_P leaks. i_{ST} is low and short-lived (by design) because M_{SW}'s high v_{GS} suppresses M_P's v_{SD} as v_I's climb collapses M_P's v_{SG}. i_G is much greater because M_N's v_{GS} and v_{DS} are high across the time M_P leaks i_{ST}. So the P_{ST} that v_{DD} supplies with i_{ST} across t_{ST} is low.

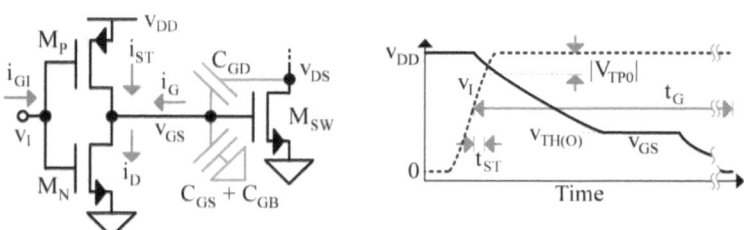

Fig. 31. Gate driver opening switch.

For v_I to rise in the first place, the pre-driver must supply the q_{GI} needed to charge M_N and M_P's gate capacitances C_{GN} and C_{GP} to v_{DD}. Since M_{SW}'s v_{GS} (which sets M_N's v_{DS}) is close to v_{DD} across this transition, v_I's climb inverts M_N into saturation. C_{CHN} in C_{GB} therefore charges across V_{TN0}, C_{OLN}'s in C_{GSN} and C_{GDN} across v_{DD}, and $(2/3)C_{CHN}$ in C_{GSN} across $v_{DD}-V_{TN0}$. Since M_N's v_{GS} matches v_{DS} at v_{DD} after v_I rises, v_{DS}'s fall charges C_{OLN} and $0.5C_{CHN}$'s average $0.25C_{CHN}$ in C_{GDN} across v_{DD}. So in all, C_{GN} requires

$$q_{GN} \approx C_{CHN}V_{TN0} + C_{OLN}\left(3v_{DD}\right) + \left(\frac{2}{3}\right)C_{CHN}\left(v_{DD} - V_{TN0}\right) + \frac{C_{CHN}v_{DD}}{4}$$

$$\approx C_{OLN}\left(3v_{DD}\right) + C_{CHN}\left(v_{DD} + \frac{V_{TN0}}{3}\right)$$

. (64)

M_P starts in triode and ends in cut off when v_I climbs because M_{SW}'s v_{GS}, which is high across v_I's transition, suppresses M_P's v_{SD}. C_{OLP}'s in C_{GSP} and C_{GDP} therefore charge across v_{DD}, $0.5C_{CHP}$'s in C_{GSP} and C_{GDP} across $v_{DD} - |V_{TP0}|$, and C_{CH} in C_{GB} across $|V_{TP0}|$. After v_I rises, as M_{SW}'s v_{GS} falls and M_P's v_{SD} climbs in cut off, C_{OLP} in C_{GDP} charges across another v_{DD}. So C_{GP} requires

$$q_{GP} \approx C_{OLP}\left(3v_{DD}\right) + \left(\frac{C_{CHP}}{2} + \frac{C_{CHP}}{2}\right)\left(v_{DD} - |V_{TP0}|\right) + C_{CHP}|V_{TP0}|$$

. (65)

$$= \left(3C_{OLP} + C_{CHP}\right)v_{DD}$$

q_{GN} and q_{GP} are the q_{GI} that v_{DD} supplies to C_{GI}. And P_{GI} is the *driver gate-charge power* that v_{DD} supplies with q_{GI}:

$$P_{GI} = v_{DD}i_{GI(AVG)} = v_{DD}\left(\frac{q_{GI}}{t_{SW}}\right) = v_{DD}\left(\frac{q_{GN} + q_{GP}}{t_{SW}}\right). \tag{66}$$

Together, C_{GI}'s q_{GI} and M_P's i_{ST} draw $P_{D(O)}$ from v_{DD} when M_{SW} opens:

$$P_{D(O)} = P_{GI} + P_{ST(O)}. \tag{67}$$

6.4. Driver Power

M_{SW}'s C_G receives charge q_G or C_Gv_{DD} to charge to v_{DD}. With this q_G, v_{DD} supplies q_Gv_{DD} or $C_Gv_{DD}^2$ and C_G stores $0.5C_Gv_{DD}^2$. This means that M_P burns half the energy v_{DD} supplies. M_N later dissipates the other half when M_N drains C_G. So v_{DD} ultimately loses the P_G that C_G draws.

v_{DD} similarly loses the P_{GI} that C_{GI} requires. v_{DD} also leaks P_{ST} when closing and opening M_{SW}. But P_{ST} is so much lower than P_G (by design) that P_D reduces to the q_G and q_{GI} that C_G and C_{GI} need:

$$P_D = P_{D(C)} + P_{D(O)}$$

$$\approx P_G + P_{GI} \approx v_{DD}\left(\frac{q_G + q_{GI}}{t_{SW}}\right) = v_{DD}\left(\frac{q_G + q_{GN} + q_{GP}}{t_{SW}}\right). \tag{68}$$

Example 14: Determine P_D and σ_D for M_{EG}'s driver in the ideal boost from Examples 7, 9, and 13.

Solution:

$$d_{DO} = 48\%, \; v_{DD} = v_O = 4 \text{ V}, \; V_{TN0} = 400 \text{ mV}, \; v_{SWO} = 4.8 \text{ V},$$

$$C_{OX}'' = 6.9 \text{ fF/µm}^2, \; L_{OL} = 30 \text{ nm}, \; C_{OL} = 10 \text{ pF},$$

$$C_{CH} = 66 \text{ pF from Examples 7 and 9}$$

$$L_{CH} = 190 \text{ nm}, \; W_N = 14 \text{ µm}, \; W_P = 26 \text{ µm from Example 13}$$

$$q_G \approx C_{OL}\left(2v_{DD} + v_{SW}\right) + C_{CH}\left(v_{DD} + \frac{V_{TN0}}{4}\right)$$

$$= 10p\left[2(4) + 4.8\right] + 66p\left(4 + \frac{400m}{4}\right)$$

$$= 400 \text{ pC}$$

$$C_{OLN} = C_{OX}''W_NL_{OL} = (6.9m)(14µ)(30n) = 2.9 \text{ fF}$$

$$C_{CHN} = C_{OX}''W_NL_{CH} = (6.9m)(14µ)(190n) = 18 \text{ fF}$$

$$q_{GN} \approx C_{OLN}\left(3v_{DD}\right) + C_{CHN}\left(v_{DD} + \frac{V_{TN0}}{3}\right)$$

$$= (2.9f)(3)(4) + (18f)\left(4 + \frac{400m}{3}\right) = 110 \text{ fC}$$

$$C_{OLP} = C_{OX}''W_PL_{OL} = (6.9m)(26µ)(30n) = 5.4 \text{ fF}$$

$$C_{CHP} = C_{OX}''W_PL_{CH} = (6.9m)(26µ)(190n) = 34 \text{ fF}$$

$$q_{GP} \approx (3C_{OLP} + C_{CHP})v_{DD} = [3(5.4f) + 24f](4) = 160 \text{ fC}$$

$$P_D \approx v_{DD}\left(\frac{q_G + q_{GN} + q_{GP}}{t_{sw}}\right)$$

$$= 4\left(\frac{400p + 110f + 160f}{1\mu}\right) = 1.6 \text{ mW}$$

$$\sigma_D = \frac{P_D}{(i_O/d_{DO})d_{EI}v_{IN}} = \frac{1.6m}{(250m/48\%)(1)(2)} = 0.2\%$$

Note: q_G is usually much greater than q_{GI}. This P_D, which is 0.2% of P_{IN}, excludes the power M_{DO}'s driver consumes.

---

7. Leaks

MOS transistors also embody *source–* and *drain–body junction capacitances* C_{SB} and C_{DB}. The terminals that do not connect to the switching nodes connect to ground, v_{IN}, or v_O, which are nearly fixed. So C_{SB} and C_{DB} add *input* and *output switch-node capacitances* C_{SWI} and C_{SWO} that charge and discharge with v_{SWI} and v_{SWO} in Fig. 32. *Electrostatic-discharge protection* (ESD), pads, pins, and board connections also add capacitance to C_{SWI} and C_{SWO}.

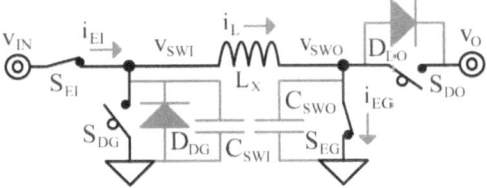

Fig. 32. Switched inductor with parasitic diodes and capacitances.

Charging C_{SWI} and C_{SWO} requires energy that v_O unfortunately does not fully recovers. Power switches also leak current when they are off, especially when they are large and hot. Although these losses are usually low, they become increasingly larger fractions of P_{IN} when P_O fades.

7.1. Input Switch-Node Capacitance

When closing, S_{EI} burns the $0.5C_{SWI}v_{DG}^2$ that C_{SWI} holds with $-v_{DG}$ and v_{IN} supplies the $v_{IN}q_{SWI}$ or $v_{IN}C_{SWI}(v_{IN} + v_{DG})$ needed to charge C_{SWI} from $-v_{DG}$ to v_{IN}. When S_{EI} opens, L_X and v_O recover the $0.5C_{SWI}v_{IN}^2$ held in C_{SWI} with v_{IN} and L_X supplies $0.5C_{SWI}v_{DG}^2$ to charge C_{SWI} to $-v_{DG}$. A t_{DT} into t_D, S_{DG} burns $0.5C_{SWI}v_{DG}^2$ to drain C_{SWI} to ground. And a t_{DT} before t_D ends, L_X loses another $0.5C_{SWI}v_{DG}^2$ to charge C_{SWI} to $-v_{DG}$. So of the energy lost, L_X and v_O recovers $0.5C_{SWI}v_{IN}^2$, which means the average *input switched-node power* P_{SWI} leaked across t_{SW} is

$$P_{SWI} = E_{SWI}f_{SW} = \left(\frac{C_{SWI}}{t_{SW}}\right)\left(2v_{DG}^2 + 0.5v_{IN}^2 + v_{IN}v_{DG}\right). \tag{69}$$

7.2. Output Switch-Node Capacitance

When S_{EG} opens, L_X supplies $0.5C_{SWO}(v_O + v_{DO})^2$ to charge C_{SWO} to $v_O + v_{DO}$. A t_{DT} into t_D, v_O recovers v_Oq_{SWO} or $v_OC_{SWO}v_{DO}$ when S_{DO} closes and discharges C_{SWO} from $v_O + v_{DO}$ to v_O. A t_{DT} before t_D ends, L_X supplies $0.5C_{SWO}[(v_O + v_{DO})^2 - v_O^2]$ to charge C_{SWO} from v_O to $v_O + v_{DO}$. Then, when t_E starts, S_{EG} burns the $0.5C_{SWO}(v_O + v_{DO})^2$ that C_{SWO} holds with $v_O + v_{DO}$. So of the energy lost, v_O only recovers $v_OC_{SWO}v_{DO}$. The average *output switched-node power* P_{SWO} leaked across t_{SW} is therefore

$$P_{SWO} = E_{SWO}f_{SW}$$

$$= \left(\frac{C_{SWO}}{t_{SW}}\right)\left[1.5(v_O + v_{DO})^2 - v_Ov_{DO} - 0.5v_O^2\right]. \tag{70}$$

$$= \left(\frac{C_{SWO}}{t_{SW}}\right)\left(v_O^2 + 1.5v_{DO}^2 + 2v_Ov_{DO}\right)$$

Example 15: Determine P_{SWI} and P_{SWO} in a synchronous buck–boost when v_{IN} is 2 V, v_O is 4 V, t_{SW} is 1 μs, C_{SWI} and C_{SWO} are 5 pF each, and D_{DG} and D_{DO} drop 400 mV.

Solution:

$$P_{SWI} = \left(\frac{C_{SWI}}{t_{SW}}\right)\left(2v_{DG}{}^2 + 0.5v_{IN}{}^2 + v_{IN}v_{DG}\right)$$

$$= \left(\frac{5p}{1\mu}\right)\left[2(400m)^2 + 0.5(2)^2 + (2)(400m)\right] = 16 \ \mu W$$

$$P_{SWO} = \left(\frac{C_{SWO}}{t_{SW}}\right)\left(v_O{}^2 + 1.5v_{DO}{}^2 + 2v_O v_{DO}\right)$$

$$= \left(\frac{5p}{1\mu}\right)\left[4^2 + 1.5(400m)^2 + 2(4)(400m)\right] = 97 \ \mu W$$

Note: C_{SWO} leaks more than C_{SWI} because L_X and v_O recover more when v_{IN} collapses than L_X recovers when v_{SWO} drops v_{DO}.

7.3. Cut-Off Power

MOS current is zero when v_{GS} and v_{DS} are zero. Raising v_{DS} when v_{GS} is zero establishes an electric field that induces some i_{DS}. Body diodes also conduct zero current when their voltages are zero and close to *reverse saturation current* I_S when they reverse. These *off currents* i_{OFF} climb with *channel width* W_{CH} and *junction temperature* T_J. So the *off resistance* R_{OFF} that they present falls with increasing W_{CH} and T_J.

Power-supply switches and body diodes usually leak noticeable i_{OFF} because they are large. Plus, the power they burn when they conduct heats them to an extent that they remain hot when they open. A 30-mm wide, 180-nm long MOSFET, for example, can leak 80 nA with 1.8 V at 25 °C and 800 nA at 125 °C. This is 0.38–3.8 TΩ per L/W square.

In a switched inductor, energize switches are off across t_D. S_{EI} drops v_{IN} and S_{EG} drops v_O across the t_D that excludes dead times. When dead-time diodes conduct, S_{EI} drops v_{IN} and v_{DG} and S_{EG} drops v_O and v_{DO}. So the average cut-off power they consume across t_{SW} is

$$P_{OFF(E)} = \left(\frac{v_{IN}^2}{R_{EI(OFF)}} + \frac{v_O^2}{R_{EG(OFF)}} \right) \left(\frac{t_D - 2t_{DT}}{t_{SW}} \right)$$

$$+ \left[\frac{\left(v_{IN} + v_{DG} \right)^2}{R_{EI(OFF)}} + \frac{\left(v_O + v_{DO} \right)^2}{R_{EG(OFF)}} \right] \left(\frac{2t_{DT}}{t_{SW}} \right). \qquad (71)$$

$$\approx \left(\frac{v_{IN}^2}{R_{EI(OFF)}} + \frac{v_O^2}{R_{EG(OFF)}} \right) \left(\frac{t_D}{t_{SW}} \right)$$

But since t_{DT}'s are small fractions of t_{SW} and diode voltages are usually lower than v_{IN} and v_O, the effects of v_{DG} and v_{DO} are minimal.

Drain switches are off across t_E and dead times within t_D. S_{DG} drops v_{IN} and S_{DO} drops v_O across t_E and S_{DG} drops v_{DG} and S_{DO} drops v_{DO} across t_{DT}'s. So the average cut-off power they consume across t_{SW} is

$$P_{OFF(D)} = \left(\frac{v_{IN}^2}{R_{DG(OFF)}} + \frac{v_O^2}{R_{DO(OFF)}} \right) \left(\frac{t_E}{t_{SW}} \right) + \left[\frac{v_{DG}^2}{R_{DG(OFF)}} + \frac{v_{DO}^2}{R_{DO(OFF)}} \right] \left(\frac{2t_{DT}}{t_{SW}} \right)$$

$$\approx \left(\frac{v_{IN}^2}{R_{DG(OFF)}} + \frac{v_O^2}{R_{DO(OFF)}} \right) \left(\frac{t_E}{t_{SW}} \right) \qquad .(72)$$

The effects of v_{DG} and v_{DO} are minimal because D_{DG} and D_{DO} conduct short t_{DT} fractions of t_{SW} and drop lower voltages than v_{IN} and v_O.

Excluding dead times, S_{EI} and S_{DG} alternate conduction to v_{SWI} and S_{EG} and S_{DO} alternate conduction to v_{SWO}. So one switch is always off at v_{SWI} and one at v_{SWO}. The one at v_{SWI} drops v_{IN} and the one at v_{SWO} drops v_O. Since off resistances for similarly large switches are comparable, *cut-off power* P_{OFF} is largely consistent and similar across time:

$$P_{OFF} \approx \left(\frac{v_{IN}^2}{R_{EI(OFF)}} + \frac{v_O^2}{R_{EG(OFF)}} \right) \left(\frac{t_D}{t_{SW}} \right) + \left(\frac{v_{IN}^2}{R_{DG(OFF)}} + \frac{v_O^2}{R_{DO(OFF)}} \right) \left(\frac{t_E}{t_{SW}} \right)$$

$$\approx \frac{V_{IN}^2}{R_{I(OFF)}} + \frac{V_O^2}{R_{O(OFF)}}. \tag{73}$$

Example 16: Determine P_{OFF} in a buck–boost when v_{IN} is 2 V, v_O is 4 V,

W's are 50 mm, L's are 250 nm, L_{OL} is 30 nm, T_J is 125 °C,

and $R_{OFF/SQ}$ at this T_J is 380 GΩ per L/W square.

Solution:

$$L_{CH} = L - 2L_{OL} = 250n - 2(30n) = 190 \text{ nm}$$

$$R_{OFF} = R_{OFF/SQ} \left(\frac{L_{CH}}{W_{CH}} \right) = 380G \left(\frac{190n}{50m} \right) = 1.4 \text{ M}\Omega$$

$$P_{OFF} \approx \frac{V_{IN}^2}{R_{OFF}} + \frac{V_O^2}{R_{OFF}} = \frac{2^2}{1.4M} + \frac{4^2}{1.4M} = 14 \text{ } \mu W$$

---

8. Design

8.1. Optimal Power Setting

Power losses normally consume the lowest fraction of input power at a particular load level. With sufficient flexibility, engineers can define and set this *optimal output power* P_O'. In practice, however, applications and technologies impose operating conditions and parametric limits that constrain P_O'. Even then, design choices can still influence P_O'.

Setting P_O' to $P_{O(MAX)}$ saves the most power, but not the most energy, especially when $P_{O(MAX)}$ is an improbable extreme. The system saves the most energy when P_O' is at the most probable setting. If this setting is unknown, halfway between $P_{O(MIN)}$ and $P_{O(MAX)}$ is often a good alternate. Although more involved and less practical, P_O' can also be the P_O that produces the highest peak efficiency or the highest average efficiency.

8.2. Power Switch

MOSFETs require power in four ways: ohmic power when they conduct i_L, gate-drive power when they close, i_{DS}–v_{DS} overlap power when they transition, and off power when they are open. Of these, P_{OFF} is usually a negligible part of P_O'. Although P_{IV} may not be as insignificant, gate drivers can reduce the impact of P_{IV} on P_O'. The only design variables that can suppress P_R and P_G are MOSFET dimensions.

Longer channels raise *channel resistance* R_{CH} and gate capacitance, and as a result, gate charge. So the R_{CH} and q_G that set P_R and P_G increase with *channel length* L_{CH}. This means that the *MOS power* P_{MOS} that P_R and P_G require is minimal when L is the *minimum allowable length* L_{MIN} that can sustain v_{SW}'s swing without breakdown effects:

$$P_{MOS} \equiv P_R + P_G . \tag{74}$$

Wider channels decrease R_{CH} and increase C_G. So the R_{CH} that sets P_R falls with increasing W_{CH}'s as the q_G that sets P_G climbs in Fig. 33:

$$P_R = i_{R(RMS)}{}^2 R_{CH} = \frac{k_R}{W_{CH}} \tag{75}$$

and
$$P_G = E_G f_{SW} = v_{DD} q_G f_{SW} = k_G W_{CH}, \tag{76}$$

where k_R and k_G are W_{CH}-independent coefficients. These opposing trends desensitize P_{MOS} from W_{CH}.

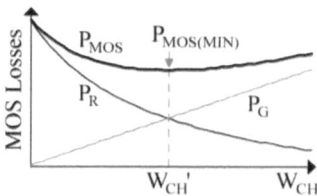

Fig. 33. Ohmic and gate-charge MOS power.

Still, P_{MOS} falls with P_R when W_{CH} is narrow and rises with P_G when W_{CH} is wide. P_{MOS} is minimal when additional charge losses in P_G cancel P_R savings. This happens when P_{MOS}'s slope $\partial P_{MOS}/\partial W_{CH}$ is zero:

$$\frac{\partial P_{MOS}}{\partial W_{CH}}\bigg|_{W_{CH}'} = \frac{\partial P_R}{\partial W_{CH}}\bigg|_{W_{CH}'} + \frac{\partial P_G}{\partial W_{CH}}\bigg|_{W_{CH}'} = -\frac{k_R}{W_{CH}'^2} + k_G = 0, \qquad (77)$$

which results at the *optimal channel width* W_{CH}':

$$W_{CH}' = \sqrt{\frac{k_R}{k_G}}, \qquad (78)$$

when P_R and P_G match:

$$P_R\big|_{W_{CH}'} = P_G\big|_{W_{CH}'} = \sqrt{k_R k_G}, \qquad (79)$$

and $P_{MOS(MIN)}$ is twice the P_R and P_G that W_{CH}' sets:

$$P_{MOS(MIN)} = P_{MOS}\big|_{W_{CH}'} = P_R\big|_{W_{CH}'} + P_G\big|_{W_{CH}'} = 2P_R\big|_{W_{CH}'} = 2\sqrt{k_R k_G}. \qquad (80)$$

In practice, on-chip, bond-wire, and board contacts and traces add resistance to switches. So ohmic power is greater than the P_R that R_{CH} establishes. Still, switches require the least power with W_{CH}'.

k_R and k_G are functions of parameters and variables that applications and process technologies frequently define. v_{IN} and v_O, for example, set duty cycle. Accuracy and noise sensitivity often specify or constrain current ripple and switching frequency. And device parameters dictate the shortest dead time that keeps adjacent switches from cross conducting. In other words, k_R and k_G are often predetermined.

Example 17: Determine M_{DG}'s optimal W, L, P_{MOS}, and σ_{MOS} in a buck–boost when v_{DD} for M_{DG}'s gate driver is v_{IN}, v_{IN} is 2 V, v_O is 4 V, D_{DG} and D_{DO} drop 400 mV, i_O is 250 mA, L_X is 10 μH, t_{SW} is 1 μs, t_{DT} = 50 ns, L_{MIN} is 250 nm, L_{OL} is 30 nm, K_N' is 200 μA/V^2, C_{OX}'' is 6.9 fF/μm^2, and V_{TN0} is 400 mV.

Solution:

$$d_{EI} = d_E = \frac{v_O}{v_{IN} + v_O} + \left(\frac{v_{DO} + v_{DG}}{v_{IN} + v_O}\right)\left(\frac{2t_{DT}}{t_C}\right)$$

$$= \frac{4}{2+4} + \left(\frac{400m+400m}{2+4} \right) \left[\frac{2(50n)}{1\mu} \right] = 68\%$$

$$\therefore \quad d_{DO} = d_D = 1 - d_E = 1 - 68\% = 32\%$$

$$\Delta i_L = \left(\frac{v_E}{L_X} \right) d_E t_{SW} \approx \left(\frac{2}{10\mu} \right) (68\%)(1\mu) = 140 \ mA$$

$$L \equiv L_{MIN} = 250 \ nm$$

$$\therefore \quad L_{CH} = L - 2L_{OL} = 250n - 2(30n) = 190 \ nm$$

$$R_{CH} \approx \left(\frac{L_{CH}}{W_{CH}} \right) \left[\frac{1}{K_N{}'(v_{DD} - V_{TN0})} \right]$$

$$= \left(\frac{190n}{W_{CH}} \right) \left[\frac{1}{(200\mu)(2 - 400m)} \right] = \frac{590\mu}{W_{CH}}$$

$$P_R = \left[\left(\frac{i_O}{d_{DO}} \right)^2 + \left(\frac{0.5\Delta i_L}{\sqrt{3}} \right)^2 \right] R_{CH} \left(d_D - \frac{2t_{DT}}{t_{SW}} \right)$$

$$= \left\{ \left(\frac{250m}{32\%} \right)^2 + \left[\frac{0.5(140m)}{\sqrt{3}} \right]^2 \right\} \left(\frac{590\mu}{W_{CH}} \right) \left[32\% - \frac{2(50n)}{1\mu} \right]$$

$$= \frac{79\mu}{W_{CH}}$$

$$C_{OL} = C_{OX}{}''W_{CH}L_{OL} = (6.9m)W_{CH}(30n) = (210p)W_{CH}$$

$$C_{CH} = C_{OX}{}''W_{CH}L_{CH} = (6.9m)W_{CH}(190n) = (1.3n)W_{CH}$$

v_{SWI} is $-v_{DG}$ before M_{EI} closes and v_{IN} after M_{EI} closes

$$\therefore \quad v_{SWI} = v_{IN} + v_{DG} = 2 + 400m = 2.4 \ V$$

$$q_G \approx C_{OL} \left(2v_{DD} + v_{SWI} \right) + C_{CH} \left(v_{DD} + \frac{V_{T0}}{4} \right)$$

$$= \left\{ 210p[2(2) + 2.4] + 1.3n \left(2 - \frac{400m}{4} \right) \right\} W_{CH}$$

$$= (3.8n)W_{CH}$$

$$P_G = v_{DD}\left(\frac{q_G}{t_{SW}}\right) \approx (2)\left[\frac{(3.8n)W_{CH}}{1\mu}\right] = (7.6m)W_{CH}$$

$$W \equiv W_{CH}' = \sqrt{\frac{k_R}{k_G}} = \sqrt{\frac{79\mu}{7.6m}} = 100 \ \text{mm}$$

$$R_{CH} = \frac{590\mu}{W_{CH}'} = 5.9 \ \text{m}\Omega$$

$$P_{MOS} \approx 2\sqrt{k_R k_G} = 2\sqrt{(79\mu)(7.6m)} = 1.6 \ \text{mW}$$

$$\sigma_{MOS} = \frac{P_{MOS}}{(i_O/d_{DO})d_{EI}v_{IN}} = \frac{1.5m}{(250m/32\%)(68\%)(2)} = 0.2\%$$

Note: P_{MOS}, which includes M_{DG}'s P_R and the driver's P_G, is much lower than P_R in Example 2 because R_{CH} is much lower.

8.3. Gate Driver

Similar transition times balance propagation delays and distribute switching losses across the switching period. This way, dead times and response time are consistent and peak transient power balances across switching events. Pull-up and -down resistances in the gate driver (which transistor W/L's, K_N' and K_P', v_{GS} and v_{SG}, and V_{TN0} and V_{TP0} set) match opposing v_{SW} transition times $t_{V(C)}$ and $t_{V(O)}$ when R_P-to-R_N's ratio is

$$\frac{R_P}{R_N} = \left(\frac{W_N}{W_P}\right)\left(\frac{L_P}{L_N}\right)\left(\frac{K_N'}{K_P'}\right)\left[\frac{v_{DD} - V_{TN0} - 0.5v_{TH}}{v_{DD} - |V_{TP0}| - 0.5(v_{DD} - v_{TH})}\right]$$

$$\equiv \left(\frac{v_{DD} - v_{TH(C)}}{v_{TH(O)}}\right)\left(\frac{C_{OL}v_{SW} + 0.25C_{CH}v_{TH(O)}}{C_{OL}v_{SW} + 0.25C_{CH}v_{TH(C)}}\right) \tag{81}$$

Although not perfectly matched, these resistances produce similar i_{DS} transitions $t_{I(C)}$ and $t_{I(O)}$. And since dead-time diodes set v_{DS} before and after switches close, v_{DS} swings the same v_{SW} when closing and opening.

So excluding reverse recovery, the overlap power that switches consume when closing and opening reduces to

$$P_{IV} = P_{I(C)}\left(\frac{t_{I(C)}}{t_{SW}}\right) + P_{V(C)}\left(\frac{t_{V(C)}}{t_{SW}}\right) + P_{I(O)}\left(\frac{t_{I(O)}}{t_{SW}}\right) + P_{V(O)}\left(\frac{t_{V(O)}}{t_{SW}}\right)$$

$$\approx \left(i_{L(LO)} + i_{L(HI)}\right)v_{SW}\left(\frac{t_I}{3t_{SW}} + \frac{t_V}{2t_{SW}}\right) \tag{82}$$

$$= 2\left(\frac{i_O}{d_{DO}}\right)v_{SW}\left(\frac{t_I}{3t_{SW}} + \frac{t_V}{2t_{SW}}\right) = \frac{k_{IV}}{W_N}$$

Switches either energize $i_{L(LO)}$ to $i_{L(HI)}$ or drain $i_{L(HI)}$ to $i_{L(LO)}$, so they steer $i_{L(LO)}$ in one transition and $i_{L(HI)}$ in the other. Since $i_{L(LO)}$ and $i_{L(HI)}$ average $i_{L(AVG)}$, their sum is twice the $i_{L(AVG)}$ that i_O and d_{DO} set. The only design variable left is the R_N that sets R_P, t_I, and t_V.

Since W_N relates to R_N and W_P to the R_P that R_N sets, R_N determines the total gate capacitance of the driver. Increasing W_N not only reduces the R_N that shortens t_I and t_V and reduces P_{IV} but also increases the C_{GI} that v_{DD} charges. So P_{GI} in the driver climbs with W_N as P_{IV} falls:

$$P_{GI} = v_{DD}\left(\frac{q_{GI}}{t_{SW}}\right) = v_{DD}\left(\frac{q_{GN} + q_{GP}}{t_{SW}}\right) = k_{GI}W_N, \tag{83}$$

where k_{IV} and k_{GI} are W_N-independent coefficients for P_{IV} and P_{GI}.

Longer channels increase channel resistance and gate capacitance, and as a result, gate charge. So the R_P, R_N, and q_{GI} that set P_{IV} and P_{GI} climb with L_{CH}. This means that P_{IV} and P_{GI}'s sum is lowest when L is the L_{MIN} that can sustain v_{DD} without breakdown effects.

P_{GI}'s rise with W_N in Fig. 34 opposes P_{IV}'s fall like P_G and P_R in P_{MOS}. So P_{IV} and P_{GI}'s sum is minimal when additional charge losses cancel P_{IV} savings. This happens when the slope of P_{IV} and P_{GI}'s sum is zero, which results when W_N is at the optimal width W_N' and P_{IV} and P_{GI} match:

$$\frac{\partial P_{IV}}{\partial W_N}\bigg|_{W_N'} + \frac{\partial P_{GI}}{\partial W_N}\bigg|_{W_N'} = -\frac{k_{IV}}{W_N'^2} + k_{GI} = 0, \tag{84}$$

$$W_N' = \sqrt{\frac{k_{IV}}{k_{GI}}}, \tag{85}$$

and

$$P_{IV}\big|_{W_N'} = P_{GI}\big|_{W_N'} = \sqrt{k_{IV}k_{GI}}. \tag{86}$$

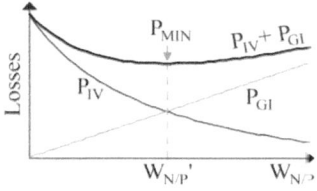

Fig. 34. Overlap and driver gate-charge power.

The W_P and W_N that raise P_{GI} to the point P_{GI} matches P_{IV} can be so high that i_P and i_N can reach i_L levels. This amplifies the effects of i_P and i_N on the closing and opening thresholds that v_{GS} reaches when v_{DS} transitions. Although incorporating this shift in v_{TH} is not impossible, the $v_{TH(C)}$ and $v_{TH(O)}$ that set P_{IV} in Section 5 are good first-order approximations. Computer models and simulations can account for the rest, including sub-threshold and weak-inversion effects.

Example 18: Determine the optimal W's and L's, P_{IV}, and P_{GI} for the gate driver that switches M_{EG} in the boost from Examples 7, 9, 10, and 13 when L_{MIN} is 250 nm and i_O is 250 mA.

Solution:

$d_E = 52\%$, $d_{DO} = d_D = 48\%$, $\Delta i_L = 100$ mA from Example 7

$v_{DD} = 4$ V, $V_{TN0} = 400$ mV, $K_N' = 200$ µA/V^2, $L_{OL} = 30$ nm,

$C_{OX}'' = 6.9$ fF/µm^2, $C_{OL} = 10$ pF, $C_{CH} = 66$ pF,

$v_{SWO} = 4.8$ V, $v_{TH(C)} = 530$ mV from Example 9

$v_{TH(O)} = 550$ mV from Example 10

$V_{TP0} = -400$ mV, $K_P' = 40$ µA/V^2 from Example 13

$L \equiv L_{MIN} = 250$ nm

$\therefore \quad L_{CH} = L - 2L_{OL} = 250n - 2(30n) = 190$ nm

$$R_P \approx \frac{L_{CH}}{W_P K_P'\left[V_{DD} - |V_{TP0}| - 0.5\left(V_{DD} - V_{TH(C)}\right)\right]}$$

$$= \frac{190n}{W_P(40\mu)\left[4 - 400m - 0.5(4 - 530m)\right]} = \frac{2.6m}{W_P}$$

$$R_N \approx \frac{L_{CH}}{W_N K_N'\left(V_{DD} - V_{TN0} - 0.5v_{TH(O)}\right)}$$

$$= \frac{190n}{W_N(200\mu)\left[4 - 400m - 0.5(550m)\right]} = \frac{290\mu}{W_N}$$

$$\frac{R_P}{R_N} \equiv \left(\frac{V_{DD} - V_{TH(C)}}{v_{TH(O)}}\right)\left(\frac{C_{OL}v_{SWO} + 0.25C_{CH}v_{TH(O)}}{C_{OL}v_{SWO} + 0.25C_{CH}v_{TH(C)}}\right)$$

$$= \left(\frac{4 - 530m}{550m}\right)\left[\frac{10p(4.8) + 0.25(66p)(550m)}{10p(4.8) + 0.25(66p)(530m)}\right] = 6.3$$

$$\rightarrow \quad \frac{R_P}{R_N} = \left(\frac{2.6m}{W_P}\right)\left(\frac{W_N}{290\mu}\right) = 6.3 \quad \therefore \quad W_P = 1.4W_N$$

$$t_V \approx \left(\frac{R_N}{v_{TH(O)}}\right)\left[C_{OL}v_{SWO} + \left(\frac{C_{CH}}{4}\right)v_{TH(O)}\right]$$

$$= \left(\frac{290\mu/W_N}{550m}\right)\left[(10p)(4.8) + \left(\frac{66p}{4}\right)(550m)\right] = \frac{30f}{W_N}$$

$$t_{I(O)} \approx R_N\left[2C_{OL} + \left(\frac{2}{3}\right)C_{CH}\right]\ln\left(\frac{v_{TH(O)}}{V_{TN0}}\right)$$

$$= \left(\frac{290\mu}{W_N}\right)\left[2(10p) + \left(\frac{2}{3}\right)(66p)\right]\ln\left(\frac{550m}{400m}\right) = \frac{5.9f}{W_N}$$

$$P_{IV} \approx 2\left(\frac{i_O}{d_{DO}}\right) v_{SWO}\left(\frac{t_I}{3t_{SW}} + \frac{t_V}{2t_{SW}}\right)$$

$$= 2\left(\frac{250m}{48\%}\right)(4.8)\left[\frac{5.9f}{3(1\mu)} + \frac{30f}{2(1\mu)}\right]\left(\frac{1}{W_N}\right) = \frac{85n}{W_N}$$

$$q_{GP} = \left(3C_{OLP} + C_{CHP}\right) v_{DD}$$

$$= C_{OX}"W_P\left(3L_{OL} + L_{CH}\right) v_{DD}$$

$$= (6.9m)(1.4W_N)\left[3(30n) + 190r\right](4) = (11n)W_N$$

$$q_{GN} = C_{OLN}\left(3v_{DD}\right) + C_{CHN}\left(v_{DD} + \frac{V_{TN0}}{3}\right)$$

$$= C_{OX}"W_N\left[L_{OL}\left(3v_{DD}\right) + L_{CH}\left(v_{DD} + \frac{V_{TN0}}{3}\right)\right]$$

$$= (6.9m)W_N\left[(30n)(3)(4) + (190n)\left(4 + \frac{400m}{3}\right)\right]$$

$$= (7.9n)W_N$$

$$P_{GI} = v_{DD}\left(\frac{q_{GP} + q_{GN}}{t_{SW}}\right) = (4)\left(\frac{11n + 7.9n}{1\mu}\right)W_N = (76m)W_N$$

$$W_N \equiv W_N' = \sqrt{\frac{k_{IV}}{k_{GI}}} = \sqrt{\frac{85n}{76m}} = 1.1 \text{ mm}$$

$W_P = 1.4W_N = 1.4(1.1m) = 1.5$ mm

$\therefore \quad P_{IV} \approx 80 \ \mu W \quad$ and $\quad P_{GI} \approx 84 \ \mu W$

Note: W_N and W_P are much wider than in Example 13 and P_{IV} and P_{GI} are much lower than P_{IV}'s in Examples 9 and 10.

8.4. Operation

A. Switch Configuration

The frequency that L_X draws energy E_L determines the power L_X outputs. In a buck–boost, L_X delivers E_L when L_X drains from $i_{L(HI)}$ to $i_{L(LO)}$:

$$P_L = E_L f_{SW} = \left(\frac{1}{2}\right) L_X \left(i_{L(HI)}{}^2 - i_{L(LO)}{}^2\right) f_{SW}. \qquad (87)$$

A buck is the same, except S_{EI} in Fig. 10 energizes L_X directly into v_O, so v_O receives power as v_{IN} energizes L_X. In other words, i_{IN} delivers v_{IN} power into v_O that L_X does not carry. v_O in a boost similarly receives v_{IN} power that L_X does not carry when S_{DO} drains L_X from v_{IN}.

In short, bucks and boosts deliver more power than L_X transfers. So for the same output power, L_X transfers less energy in bucks and boosts than in buck–boosts. And less energy translates to lower i_L, which burns lower ohmic, dead-time, and overlap power. These *direct transfers* is one reason why bucks and boosts are more efficient than buck–boosts.

The other reason is fewer switches. Buck–boosts need two more switches, which require additional ohmic, gate-drive, dead-time, and overlap power. Engineers should therefore resort to buck–boosts only when v_{IN}'s and v_O's operating ranges overlap.

When a buck–boost is necessary, the controller saves energy by only switching the output switches when bucking and the input switches when boosting. For this, S_{DO} should remain closed and S_{EG} open when bucking and S_{EI} closed and S_{DG} open when boosting. The controller should switch all transistors only when v_{IN} and v_O are close. This way, by switching fewer times, gate-drive losses are lower.

B. Discontinuous Conduction

In discontinuous conduction, i_L rises to $i_{L(PK)}$ and falls to zero across t_C before t_{SW} ends. The average ohmic power $P_{R(C)}$ that a switch consumes across t_C is a squared RMS translation of the current R_{CH} carries across t_E or t_D, which is a root-three reflection of $i_{L(PK)}$:

$$P_{R(C)} \approx i_{E/D(RMS)}{}^2 R_{CH} = \left(\frac{i_{L(PK)}}{\sqrt{3}}\right)^2 R_{CH}\left(\frac{t_{E/D}}{t_C}\right) = \frac{k_{RC}}{W_{CH}}. \tag{88}$$

The average gate-charge power $P_{G(C)}$ needed across t_C is the power v_{DD} supplies when feeding gate charge q_G into the gate:

$$P_{G(C)} = v_{DD}i_{G(AVG)} = v_{DD}\left(\frac{q_G}{t_C}\right) = k_{GC}W_{CH}. \tag{89}$$

Since R_{CH} and $P_{R(C)}$ fall and q_G and $P_{G(C)}$ rise with larger W_{CH}'s, $P_{R(C)}$ and $P_{G(C)}$'s sum $P_{MOS(C)}$ is minimal when the slope is zero and W_{CH} is at the optimal width W_{CH}':

$$\left.\frac{\partial P_{MOS(C)}}{\partial W_{CH}}\right|_{W_{CH}'} = \left.\frac{\partial P_{R(C)}}{\partial W_{CH}}\right|_{W_{CH}'} + \left.\frac{\partial P_{G(C)}}{\partial W_{CH}}\right|_{W_{CH}'} = -\frac{k_{RC}}{W_{CH}{}'^2} + k_{GC} = 0 \tag{90}$$

and

$$W_{CH} \equiv W_{CH}' = \sqrt{\frac{k_{RC}}{k_{GC}}}. \tag{91}$$

With this W_{CH}, $P_{R(C)}$ and $P_{G(C)}$ match, which means $P_{MOS(C)}$ reduces to

$$P_{MOS(C)} = P_{R(C)} + P_{G(C)} = 2\sqrt{k_{RC}k_{CC}}. \tag{92}$$

So across t_{SW}, the optimal switch consumes $P_{MOS(C)}$ a t_C fraction of t_{SW}:

$$P_{MOS} = P_{MOS(C)}\left(\frac{t_C}{t_{SW}}\right) = 2\left(\sqrt{k_{RC}k_{GC}}\right)t_Cf_{SW}. \tag{93}$$

Across each t_{SW}, P_O outputs the energy L_X collects with $i_{L(PK)}$ and the additional power $P_{E/D}$ that v_{IN} supplies in bucks and boosts. This $P_{E/D}$ is what v_O receives with i_L's average $0.5i_{L(PK)}$ across t_E or t_D. The resulting P_O climbs with *switching frequency* f_{SW}:

$$P_O \approx \frac{E_L}{t_{SW}} + P_{E/D}\left(\frac{t_{E/D}}{t_{SW}}\right) \approx \left[\left(\frac{1}{2}\right)L_Xi_{L(PK)}{}^2 + \left(\frac{i_{L(PK)}}{2}\right)v_Ot_{E/D}\right]f_{SW}. \tag{94}$$

This is fortunate because P_{MOS}, P_{DT}, P_{IV}, P_{GI}, and P_{SW} also scale with f_{SW}.

So if L_X's energy E_L is constant (with fixed t_E, $i_{L(PK)}$, t_D, and t_C), power-conversion efficiency would be similarly independent of P_O:

$$\eta_C = \frac{P_O}{P_{IN}} \approx \frac{P_O}{P_O + P_{MOS} + P_{DT} + P_{IV} + P_{GI} + P_{SW}} \propto \frac{f_{SW}}{f_{SW}} \neq f(P_O). \quad (95)$$

And with W_{CH}', this η_C would also be optimally high.

For this, the controller should adjust t_{SW} (not d_E), which is a form of *frequency modulation* (FM). *Constant on-time control* and *pulse-FM* (PFM), for example, fix t_E and vary the frequency that L_X delivers energy packets. *Burst mode* is a variation that adjusts either the number of consecutive energy packets delivered between conduction gaps or the conduction gap between consecutive energy packets.

Figure 35 shows the efficiency of a photovoltaic battery-charging voltage regulator that adjusts the frequency of energy packets in discontinuous conduction. η_C is nearly constant at 95% when L_X is $3 \times 3 \times 1.5$ mm^3. η_C is lower, but still constant when L_X is $1.6 \times 0.8 \times 0.8$ mm^3 because a smaller L_X is more resistive and therefore lossier. η_C drops when the *load power* P_{LD} that sets P_O nears zero because the controller consumes quiescent that does not scale with f_{SW} (or P_O).

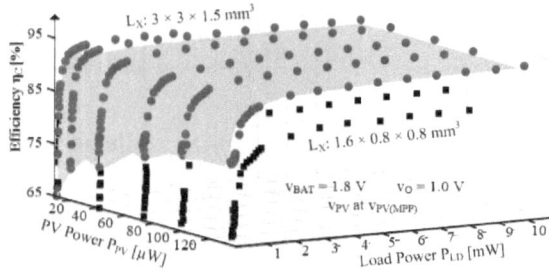

Fig. 35. Efficiency of frequency-modulated DCM system.

8.5. Power-Conversion Efficiency

Power-conversion efficiency refers to the fraction of P_{IN} that P_O outputs. η_C is ultimately a reflection of fractional losses. So increasing η_C amounts to reducing σ_{LOSS}.

A. Discontinuous Conduction

In DCM, P_R scales with $i_{L(PK)}{}^2$, t_C, and f_{SW}; P_{DT} and P_{IV} with $i_{L(PK)}$ and f_{SW}; and P_G and P_{SW} with f_{SW}. P_{CNTRL} and P_{OFF} are largely constant with respect to $i_{L(PK)}$ and f_{SW}. And P_{IN} scales with the i_O that sets P_O.

Frequency Modulation: With FM, $i_{L(PK)}$ and t_C are constant and f_{SW} scales with i_O. This way, P_{CNTRL} and P_{OFF} scale with $i_O{}^0$ and P_R, P_{DT}, P_{IV}, P_G, P_{GI}, P_{SW}, and P_{IN} with $i_O{}^1$. So their fractional losses σ_{DCM0} and σ_{DCM1} scale with $i_O{}^{-1}$ and $i_O{}^0$. In short, σ_{DCM0} falls with i_O and σ_{DCM1} is constant:

$$\sigma_{DCM0} \approx \frac{P_{CNTRL} + P_{OFF}}{P_{IN}} \propto \frac{i_O{}^0}{i_O{}^1} \propto \frac{1}{i_O} \tag{96}$$

$$\sigma_{DCM1} \approx \frac{P_R + P_{DT} + P_{IV} + P_G + P_{GI} + P_{SW}}{P_{IN}} \propto \frac{i_O{}^1}{i_O{}^1} \neq f(i_O). \tag{97}$$

When lightly loaded, i_O-dependent losses are so low that P_{CNTRL} and P_{OFF} dominate. In this region in Fig. 36, η_C rises because their σ_{DCM0} falls with i_O. η_C eventually peaks and flattens in Region II when P_R, P_{DT}, P_{IV}, P_G, P_{GI}, and P_{SW} dominate because their σ_{DCM1} is constant.

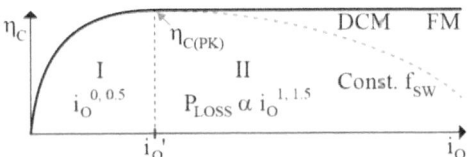

Fig. 36. Power-conversion efficiency in DCM.

Peak Modulation: When f_{SW} is constant, $i_{L(PK)}$ and t_C scale with $\sqrt{i_O}$. This way, P_G, P_{GI}, P_{SW}, P_{CNTRL}, and P_{OFF} scale with $i_O{}^0$, P_{DT} and P_{IV} with $i_O{}^{0.5}$, and P_R with $i_O{}^{1.5}$. So their σ_{DCM0}, $\sigma_{DCM0.5}$, and $\sigma_{DCM1.5}$ scale with $i_O{}^{-1}$, $i_O{}^{-0.5}$, and $i_O{}^{0.5}$. In short, σ_{DCM0} and $\sigma_{DCM0.5}$ fall and $\sigma_{DCM1.5}$ rises with i_O:

$$\sigma_{DCM0} \approx \frac{P_G + P_{GI} + P_{SW} + P_{CNTRL} + P_{OFF}}{P_{IN}} = k_{D0}\left(\frac{i_o^{\ 0}}{i_o^{\ 1}}\right) = \frac{k_{D0}}{i_o}, \qquad (98)$$

$$\sigma_{DCM0.5} \approx \frac{P_{DT} + P_{IV}}{P_{IN}} = k_{D0.5}\left(\frac{i_o^{\ 0.5}}{i_o^{\ 1}}\right) = \frac{k_{D0.5}}{i_o^{\ 0.5}}, \qquad (99)$$

$$\sigma_{DCM1.5} \approx \frac{P_R}{P_{IN}} = k_{D1.5}\left(\frac{i_o^{\ 1.5}}{i_o^{\ 1}}\right) = k_{D1.5}\sqrt{i_o}. \qquad (100)$$

When lightly loaded, i_o reduces P_R to an extent that P_{DT}, P_{IV}, P_G, P_{GI}, P_{SW}, P_{CNTRL}, and P_{OFF} dominate. η_C rises in Region I in Fig. 36 because their $\sigma_{DCM0.5}$ and σ_{DCM0} fall with i_o. η_C falls in Region II when P_R dominates because its $\sigma_{DCM1.5}$ climbs with i_o.

Their combined σ_{LOSS} reaches its minimum when its slope flattens with respect to i_o. This happens when $\sigma_{CCM1.5}$'s rise outpaces σ_{DCM0} and $\sigma_{DCM0.5}$'s fall. At this optimal $i_o{}'$, η_C maxes with one minus this σ_{LOSS} at $i_o{}'$:

$$\left.\frac{\partial\sigma_{DCM0}}{\partial i_o} + \frac{\partial\sigma_{DCM0.5}}{\partial i_o} + \frac{\partial\sigma_{DCM1.5}}{\partial i_o}\right|_{i_o{}'} = -\frac{k_{D0}}{i_o{}'^2} - \frac{k_{D0.5}}{2\sqrt{i_o{}'^3}} + \frac{k_{D1.5}}{2\sqrt{i_o{}'}} = 0. \quad (101)$$

So η_C peaks when P_R balances P_{DT}, P_{IV}, P_G, P_{GI}, P_{SW}, P_{CNTRL}, and P_{OFF}.

B. Continuous Conduction

In CCM, $P_{R(AC)}$, P_G, P_{GI}, P_{SW}, P_{CNTRL}, and P_{OFF} scale with i_o^0; P_{DT}, P_{IV}, and P_{IN} with i_o^1; and $P_{R(DC)}$ with i_o^2. So their fractional losses σ_{CCM0}, σ_{DCM1}, and σ_{DCM2} scale with i_o^{-1}, i_o^0, and i_o^1. In other words, σ_{DCM0} falls and σ_{DCM2} rises with i_o and σ_{DCM1} is constant:

$$\sigma_{CCM0} \approx \frac{P_{R(AC)} + P_G + P_{GI} + P_{SW} + P_{CNTRL} + P_{OFF}}{P_{IN}} = \frac{k_{C0}}{i_o}, \qquad (102)$$

$$\sigma_{CCM1} \approx \frac{P_{DT} + P_{IV}}{P_{IN}} = k_{C1}\left(\frac{i_o}{i_o}\right) = k_{C1}, \qquad (103)$$

$$\sigma_{CCM2} \approx \frac{P_{R(DC)}}{P_{IN}} = k_{C2}\left(\frac{i_o^{\ 2}}{i_o}\right) = k_{C2}i_o. \qquad (104)$$

When i_O is low, $P_{R(AC)}$, P_G, P_{GI}, P_{SW}, P_{CNTRL}, and P_{OFF} often outweigh i_O-dependent losses. η_C rises in Region III in Fig. 37 because their σ_{DCM0} falls with i_O. η_C flattens in Region IV when P_{DT} and P_{IV} dominate because their σ_{DCM1} is insensitive to i_O. η_C falls in Region V when $P_{R(DC)}$ dominates because its σ_{DCM2} climbs with i_O.

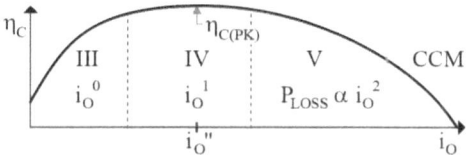

Fig. 37. Power-conversion efficiency in CCM.

Their combined σ_{LOSS} reaches its minimum when its slope flattens with respect to i_O. This happens when σ_{CCM2}'s rise outpaces σ_{DCM0}'s fall. At this optimal i_O'', η_C maxes with one minus this σ_{LOSS} at i_O'':

$$\frac{\partial \sigma_{CCM0}}{\partial i_O} + \frac{\partial \sigma_{CCM1}}{\partial i_O} + \frac{\partial \sigma_{CCM2}}{\partial i_O}\bigg|_{i_O''} = -\frac{k_{C0}}{i_O''^2} + 0 + k_{C2} = 0. \tag{105}$$

$$i_O'' = \sqrt{\frac{k_{C0}}{k_{C2}}}, \tag{106}$$

$$\sigma_{CCM1}\big|_{i_O''} = \sigma_{CCM2}\big|_{i_O''} = \sqrt{k_{C0}k_{C2}}, \tag{107}$$

$$\eta_{C(PK)} = 1 - \sigma_{CCM0}\big|_{i_O''} - \sigma_{CCM1}\big|_{i_O''} - \sigma_{CCM2}\big|_{i_O''} = 1 - \sigma_{CCM1} - 2\sqrt{k_{C0}k_{C2}}. \tag{108}$$

So η_C peaks when $P_{R(DC)}$ balances $P_{R(AC)}$, P_G, P_{GI}, P_{SW}, P_{CNTRL}, and P_{OFF}.

C. Response

The operating range of a load can sometimes exclude η_C's outer regions. Portable products, for example, can exclude Regions IV–V because loads may not reach that high. Stationary and higher-power applications, on the other hand, can exclude Regions I–II because loads are never that low.

Although not often the case, η_C can peak twice when f_{SW} is constant: once in DCM and another time in CCM. η_C can also shrink or skip regions.

η_C can, for example, reduce or jump Regions II–III when P_{DT} and P_{IV} are heavy or Regions III–IV when P_{DT} and P_{IV} are light and P_R is heavy.

Maxing η_C is ultimately more important than peaking η_C. In other words, reducing losses is more important than peaking η_C. This is because the highest η_C for a particular i_O is not necessarily where η_C peaks.

9. Summary

Switched-inductor power supplies are popular in electronics because they condition and transfer most of the power they receive. Power-conversion efficiency is high because losses are low. Still, resistances, diodes, and transistors consume power that does not reach the output.

Resistances in switches, inductors, and capacitors burn ohmic power when they conduct. The current that sets this power decomposes into static and alternating components. The dc portion is ultimately a reflection of the load. The ac portion accounts for ripples, including those that connecting and disconnecting the load creates.

Diodes that conduct dead-time current when switches are off consume power. These diodes conduct twice every switching cycle in continuous conduction. In discontinuous conduction, they only conduct once per cycle because inductor current is zero after L_X drains.

Transistors burn i_{DS}–v_{DS} overlap power when they switch. Input, output, and diode voltages set the voltage they swing and inductor current sets the current they conduct. Recovering in-transit charge held in diodes increases the current and time they conduct. Although gate drivers ensure these transitions are short, overlap power is not always low.

Luckily, diode voltages are usually so much lower than input and output voltages that transitioning across them burns a small fraction of drawn input power. Overlap power is similarly low when inductor current reaches zero in discontinuous conduction. Soft-switching events like these are desirable in power supplies.

Gate drivers and pre-drivers draw and burn the supply power needed to switch transistors: half when charging gates and the other half when draining gates. They also leak shoot-through power. This leakage is low, however, when input and output voltage transitions do not overlap.

Switch-node capacitances leak the energy they need to charge. Although the inductor and the output recover some of this energy, switches ultimately burn most of it. Large power transistors also leak current in cut off, especially when they are hot.

Although not always known, minimizing losses at the most probable load level saves the most energy. Ohmic and gate-charge power in MOSFETs can balance at this load setting with a particular channel width. Gate drivers can similarly balance driver gate-charge and i_{DS}–v_{DS} overlap power with particular N- and P-channel widths.

Since bucks steer input power into the output when they energize and boosts when they drain, bucks and boosts deliver more power than the inductor transfers. So bucks and boosts need less current to transfer power than buck–boosts. Bucks and boosts also need fewer switches. And with lower current and fewer switches, ohmic losses are lower.

Efficiency is usually low with light loads because quiescent power in the controller is a large fraction of the input power drawn. Efficiency is also low with heavy loads because dc ohmic losses scale faster with output power than input power does. Frequency modulation in discontinuous conduction steadies efficiency at a high level when MOS ohmic and gate-charge losses balance. Efficiency similarly peaks in continuous conduction when ac ripple and charge losses balance dc ohmic losses.

High efficiency is important in voltage regulators, LED drivers, and chargers because it saves energy. Although not to the same extent, it is also important in energy harvesters. Output power is more important in harvesters because ambient energy is not always available. Still, output power is higher when efficiency peaks at the maximum-power point, which is not always possible.

www.ingramcontent.com/pod-product-compliance
Lightning Source LLC
Chambersburg PA
CBHW021501210526
45463CB00002B/832